"The three women of Saanich Organics are truly inspiring. Their enthusiasm about growing food and their capacity for sharing and openness has proven to be a successful business strategy and has also developed a healthy farming community around them. I am in awe and amazed by these fabulous women." —Chrystal Bryson

"Delivery day! When I open the Saanich Organics box I am almost ravished at the sight and smell and feel of the vegetables. I am just so grateful to have farmers in my local region who have the knowledge and desire to grow such remarkable plants. And I'd ask—how can we support the growth of this phenomenon?" —Christine St. Peter, PHD, Professor, Department of Women's Studies, University of Victoria

"I remember a few times last year when I was having those bad farming moments where I would question everything I was doing. Usually those times I had wallowed in muck all day, and I was wondering why I didn't have 'a real job.' Then I would drop off my produce at Saanich Organics and I would come away feeling good and think what I am doing is a good choice and worthwhile. Hooray for Saanich Organics!" —Jana McLaughlin

"Last summer when I had harvested more bush beans than I could sell prior to leaving for a few days, I called Heather to see if Saanich Organics needed any more beans. She said they had lots, but would buy them, cut them up and freeze them for next winter's box program. That kind of impromptu support during a busy August was wonderful. Thank you!" —Randy Pearson

"After three seasons farming beside Robin, Rachel, and Heather, I had the amazing feeling of having experienced the real meaning of a farmer community. I farmed on my own before and will never go back to being isolated after experiencing the potential of farming surrounded by so much support, knowledge and cheers." — Melanie Sylvestre

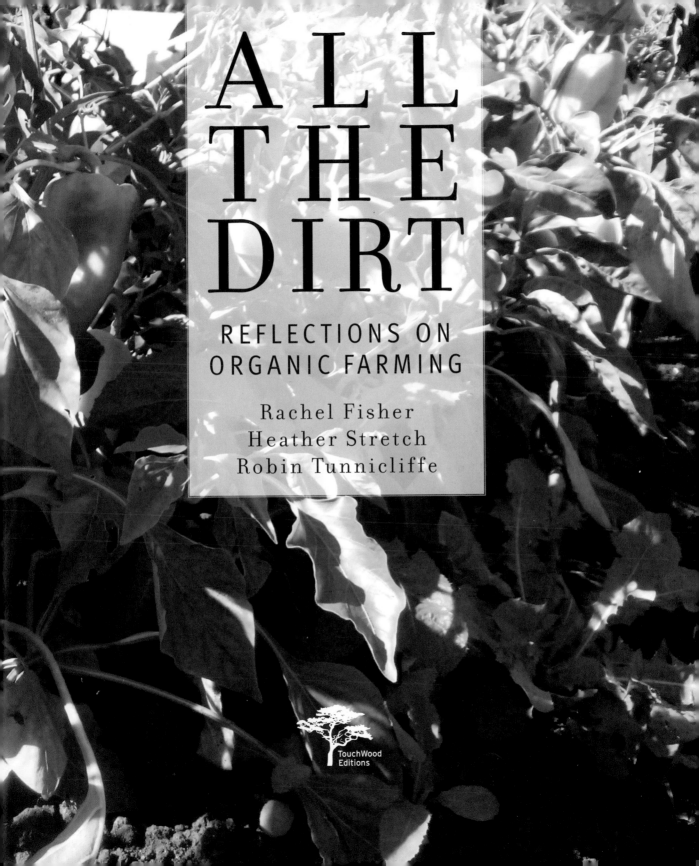

ALL THE DIRT

REFLECTIONS ON ORGANIC FARMING

Rachel Fisher
Heather Stretch
Robin Tunnicliffe

TouchWood
Editions

TouchWood Editions
touchwoodeditions.com

LIBRARY AND ARCHIVES CANADA CATALOGUING IN PUBLICATION
Fisher, Rachel, 1969–
All the dirt / Rachel Fisher, Heather Stretch, Robin Tunnicliffe.

Includes bibliographical references and index.
Also issued in electronic format.
ISBN 978-1-972129-12-8

1. Agriculture—Canada. 2. Agriculture, Cooperative—Canada.
3. Farmers—Canada—Biography. 4. Farm life—Canada. I. Stretch,
Heather, 1972– II. Tunnicliffe, Robin, 1974– III. Title.

S501.2.F58 2012   630   C2011-907093-6

Editor: Marlyn Horsdal
Proofreader: Holland Gidney
Design: Pete Kohut
Front cover photos: Tomatoes courtesy of Jill Banting, the rest courtesy of the authors.
Back cover photos: Turnip Power courtesy of Jill Banting, the rest courtesy of the authors.
Author photo on the back flap: courtesy of Madeleine Gauthier.

We gratefully acknowledge the financial support for our publishing activities from the Government of Canada through the Canada Book Fund, Canada Council for the Arts, and the province of British Columbia through the British Columbia Arts Council and the Book Publishing Tax Credit.

This book was produced using FSC®-certified, acid-free paper, processed chlorine free and printed with soya-based inks.

1 2 3 4 5 16 15 14 13 12

PRINTED IN CHINA

We would like to dedicate this book to our mentors: Tina Baynes, Rebecca Jehn, and Mary Alice Johnson, for showing us the way and leading by example. We have always known we were standing on your shoulders, and we are grateful. We'd also like to acknowledge our long-standing box customers, market regulars, and farm customers. Thanks for sharing the joy of our bounty—you make it possible.

I would like to dedicate this book to all the people who have made the farm what it is today. First and foremost, to my parents for believing in me and for being my first farmhands—the best I'll ever have. To all the friends, volunteers, and farmhands who have come out and lent a hand, you have enriched my life and continue to be sources of inspiration. —Robin

I would like to thank my parents for their love and support, and my Aunt Jane and Uncle Brian for their pivotal role in making Northbrook Farm a reality. I dedicate this book to Lamont, without whom this farm, this family, and this life would not be possible. —Heather

I would like to dedicate this book to my dad, who has always been supportive even though I didn't become a doctor or a lawyer. To my mom, who understands my passion and helps in so many large and small ways. To my partner, Grant, who is the solid ground when everything else is in flux. To my kids, who have enriched my life in countless ways. Above all, though, this book is for the Earth, who just keeps on giving and giving. —Rachel

Contents

Contents

Preface

We wrote this book because we want to encourage prospective farmers, because we love what we do, and because there is such a dire need for regenerative agriculture. We believe passionately in the need for sustainable, ethically produced food. At the same time, we don't want to delude people and lead them into farming under false pretenses. We want to present a balanced perspective.

Who We Are

We are Robin Tunnicliffe of Feisty Field Organic Farm, Heather Stretch of Northbrook Farm, and Rachel Fisher of Three Oaks Farm. We are all in our thirties, have been farming for ten years or more and together we own Saanich Organics, the business that sells our produce along with the produce from several of our friends' farms.

Why This Book

This book has been an ongoing project over the past several winters. One of the first questions we asked each other was, "Why are we writing this book?" We each had a different answer.

Rachel: When I started farming, I needed this book. I needed to know that other people were doing small-scale, sustainable agriculture, and making a living at it. I also needed to know the nuts and bolts of how they did it. How the heck do you choose a site, plan irrigation, and put up a greenhouse?

Robin: When I was gathering information to start farming, I wanted a slice of life—I wanted to see what the farming life was going to look like in the long term. I wanted numbers and I wanted facts.

Heather: I want to help people make informed choices. When I was starting out I read too many books on market gardening that promised huge yields and huge incomes with little work. I knew that if it were that easy, more people would be doing it.

There is a growing interest right now in how our food is produced, and we feel that underlying this curiosity is an unmet need for a connection to the land. People even take time off from their jobs and travel to visit our farms, volunteer to work with us, and learn what we do. People ask us all the time about our work, about what we do, and what the experience is like. "Gosh, it would take a book to answer that question," we think. Well, here is that book.

We think the best way to learn about farming is to hear farmers' stories. Each farm is unique, and each story is informative in its own practical context. We want to share with you the stories of three successful farms and to give you an honest, straight-up "this is how it is." When visitors come to our farms, depending on the day, we may tell them it's the best life in the world, or that it's the worst life. This book is an opportunity for our sober reflection, to balance the good and the bad, and, we hope, to conclude that farming truly is a worthwhile endeavour.

Introduction

Three Oaks apprentices out in full force on a pepper harvest day in 2010.

Many of us have a stereotype of a farmer as man alone on a tractor, the same tractor his father and grandfather drove before him. We invite you to let go of the idea of farming as a lifelong, multi-generational activity. Farming can take as many forms as your imagination allows—part-time, full-time, seasonal, temporary, or, of course, the lifelong passion. With a few notable exceptions, the people in our community who have continued to farm for many years own their land, and have outside capital, either from a past career or a spouse employed off-farm. They are rooted in this geographic region and are connected to the social fabric around them.

If you think you may be the lifelong-passion type of farmer but don't have these circumstances already, do not despair: none of us did either when we started. This only means that you will have to be exceptional. You can be. You are indeed up against some heavy odds, and you need to be prepared for some challenges and discouragements. But let's face it: in this culture, which doesn't value food or farmers as it should, even if you have all the advantages behind you, you will still have to be exceptional. Farming is indeed an unusual career choice.

We hope that by reading this book you will get a better idea if farming is for you.

If it is, we hope you will gain some tools to enable you to succeed. You will undoubtedly develop your own idea of what success is on your farm. For us, success means growing abundant, healthy food in healthy soil, feeling fulfilled by our work and connected to our farming community. Success also involves making enough money to provide a modest living and allow us gradually to save some money.

We were all friends before we went into business together. We met through farm community events and later joined in weekly work parties. Those work parties offered moral support as well as helping the host to tackle the big jobs that she or he didn't want to do alone. While we worked, we shared information about growing, marketing, and community events. On the most practical level, we learned from observing other people's crops and techniques. On a deeper level, coming together provided validation of our life's work that could only come from other farmers.

Deconstructing Dinner

Robin spoke at a conference on the topic of starting a farm and her talk was picked up for the radio program *Deconstructing Dinner*. It has been aired repeatedly, and we've received feedback on it from as far away as Australia. Here's a link so you can listen: www.kootenaycoopradio.com/deconstructingdinner/022609.htm

Rachel's son, Elias, helping harvest cabbage when he was three.

Gradually, the added responsibilities of families and staff meant the end of the official weekly work parties, but since we have come together to run Saanich Organics, we work more closely than ever. Farming is certainly not a lonely, isolated activity for any of us. Instead, we continue to be each other's closest friends as well as business partners.

The transition to becoming farmers was an intentional diversion from mainstream careers. We felt that the larger society around us just didn't understand what we were doing or why. This feeling came from many little experiences. At the market, most customers were great, but there were also two kinds of reactions that bothered us. Sometimes we sensed a patronizing attitude from people who seemed to think what we were doing was cute or quaint. They assumed it was a hobby, or a backyard pastime. Others expressed outrage at our prices, clearly indicating that they had no idea how much work had gone into growing the vegetables.

Robin felt disconnected from her urban friends after moving to her farm; their lives had suddenly taken such different paths. Her old friends seemed to have so much disposable income, whereas both money and time had taken on a whole new importance for her. All of a sudden, a leisurely brunch had become an impossibility; there were just so many more important ways for her to spend that money and that time. It became increasingly clear that a packet of seed was more valuable than a pint of beer.

Being with other farmers, whether at work parties, meetings, or conferences, is reaffirming. In contrast to our feeling of alienation from the mainstream, we all enjoy the kinship we feel when we are in the same room as other organic farmers. We know that we share more than an occupation. We share a common outlook, values, compassion, and an underlying love of the earth. Organic farmers are people who have taken action, and live what they believe. This is not to say we are a homogeneous group. Most in our larger community grow food, but some are also more active politically, lobbying for policy changes or running for office. Others are passionate about seed-saving and preserving heritage varieties. Some work on urban food-security issues, while others educate consumers in remote rural areas. The common thread is that we are all working in solidarity to create an alternative food system that can sustain both our planet and ourselves. Reading about the sustainability movement and hearing inspiring stories (and learning of the horrors of the industrial system) gave wings to our shovels, and made the long days feel like a contribution to something bigger than our own farms.

This book begins with our personal accounts of how we each started our farms and the experiences we garnered in subsequent years. Looking back on the early years, we've all thought, "If I'd known then what I know now . . ." These are our stories, with the benefit of hindsight. We were concerned that reading three start-up stories of similar-scale farms would be repetitive, but instead they came out as different as our personalities. They show that no two situations are the same and that there is no single right way to start out. The beauty of farming is that a creative process emerges from the interaction of your personality, your work, and your circumstances. Each farm is an expression of the farmer who created it. When we walk across a farm, we feel the essence of the farmer who works that land. Heather's chapter is packed with practical advice, Rachel's evokes the spiritual philosophy that guides her worldview, and Robin's is full of energy and humour. Then comes a chapter that delves into the ecological, social, and political issues behind our commitment to farming, followed by a collectively written chapter about our marketing business, Saanich Organics. We hope this section will provide inspiration and ideas about how others can co-operate, because it works so well for us.

Cast of Characters

Heather Stretch: owner of Northbrook Farm and co-owner of Saanich Organics. The first administrator of Saanich Organics.

Lamont Leatherman: Heather's husband. Geologist and part-time farmer.

Jackson, Walker, and Levi: Heather and Lamont's children.

Brian and Jane Stretch: Heather's uncle and aunt, co-owners (with Heather and Lamont) of the property where Northbrook Farm, Square Root Organic Farm, RiseUp Organic Farm, and L.J. are located. They also own and operate Cotyledon Farm (cut flowers and perennials) on this property.

Rachel Fisher: owner of Three Oaks Farm and co-owner of Saanich Organics. L.J. crop coordinator.

Grant Marshall: Rachel's partner.

Elias and Jade: Rachel and Grant's children.

Robin Tunnicliffe: owner of Feisty Field Organic Farm and co-owner of Saanich Organics. Media contact and newsletter writer for Saanich Organics.

Andrew Stordy: Robin's former partner who farmed with Saanich Organics for many years and remains a friend of the farm.

Chrystal Bryson: co-owner of Square Root Organic Farm (sells to Saanich Organics) and part-time administrator of Saanich Organics.

Ilya Amhrein: co-owner of Square Root Organic Farm, Chrystal's life partner and farming partner.

Melanie Sylvestre: part-time field- and greenhouse-crop manager for L.J. (Saanich Organics). Owner of RiseUp Organics (sells to Saanich Organics).

L.J.: originally, "Long John" was our nickname for the greenhouse that Saanich Organics built on an unused corner of Northbrook Farm property. Gradually, the collective Saanich Organics crops have expanded out of the greenhouse, so the name L.J. now refers to all the crop areas that we grow collectively.

Tim Deighton: Saanich Organics delivery guy extraordinaire.

Tina Baynes: original Saanich Organics co-owner and mentor to all of us, especially Robin and Lamont. Owner of Corner Farm.

Rebecca Jehn: original Saanich Organics co-owner, passionate seed-saver, and mentor to all of us. Owner of Rebecca's Garden.

Arugula is a favourite of our chefs and our market customers.

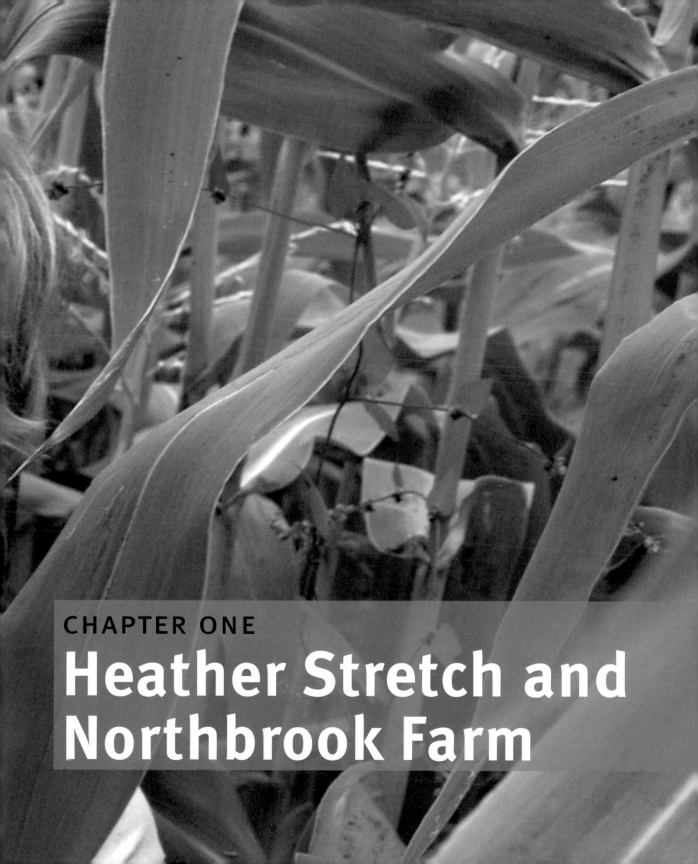

CHAPTER ONE

Heather Stretch and Northbrook Farm

Northbrook Farm is a seven-acre certified organic farm that is part of a larger piece of land co-owned by my husband Lamont and me, and my Aunt Jane and Uncle Brian. But Northbrook Farm is more than that. It is my living, my food source, my identity in the community, my sons' playground, my joy, my frustration, my business.

Why Farm?

Most people would assume that someone who starts an organic farm had a pre-existing passion for growing. In my case, that came later. My uncle offered me land to farm several years ago, and until that moment, I had never considered farming. From the moment he mentioned it, though, it seemed like a perfect fit for me. I had finished university, where I had completed an English degree, and then I had spent some time travelling and working at various jobs. All the while I had pondered what I would do for the long term.

I knew that I loved working outdoors, and that I liked physical work. I sure didn't relish the idea of sitting behind a desk all day and then having to go to a gym to stay healthy. I wanted work that would involve both my body and my brain. I was also becoming increasingly aware of the basic connection between what I put into my body and how I felt. I wanted to find work that could include a family, so that my "work life" and "family life" wouldn't be separate categories. Coupled with all this was a desire to find work that would leave a small environmental footprint. I wanted work that would benefit my own health, as well as that of my community and my planet. This seemed like a very tall order until I considered organic farming!

I accepted my uncle's offer, but deferred the move for a couple years while I took care of some other goals. In the meantime, I met Lamont, the man who was to become my husband. The poor guy fell in love with a woman who already had a plan. He decided he could go along with my plan, and moved across the continent from North Carolina to pursue my dream with me. I had land and I had my new little family of two; all I needed now was to learn how to grow vegetables, or so I thought. As time went on, I realized that I also needed to learn how to plan a farm, how to decide what to grow, how to nurture the soil, how to keep records, how to become certified organic, how to sell, and how to run a business!

Sharing Land—Heather

Recently, Robin said to me, "Did you know that there are seven distinct farm operations on this property? Strange as it may sound, I had never counted them, because each arrangement just evolved on its own. First, Brian and Jane invited us to buy a share in their land, so we began Northbrook Farm (organic veggies and berries), and they started Cotyledon Farm (cut flowers and ornamentals). A few years later, Robin, Rachel, and I decided Saanich Organics should put up a greenhouse. It only made sense to build it here, because at that time Robin and Rachel were both leasing. Next, Robin started farming a quarter-acre here. This was simply because she had reached capacity at her Feisty Field location, and Lamont and I were not using our whole field. Around this time, I began the pattern of providing land for former apprentices to start growing their own crops. We had a field that was in hay, which a neighbour cut once a year. So, gradually, parts of that hay field were transformed into Square Root Organic Farm (Chrystal and Ilya) and RiseUp Organics (Melanie). Last but not least, several years ago Rebecca approached me because she had a problem. She has a seed business, but had to move farms. In order to maintain the certified organic status of the seeds she sells, she wanted to grow them on my certified organic farm rather than her new farm, which was in transition to certified organic. She then learned that some of her seed crops do better here on my warm, dry slope than they do on her home farm, so Rebecca's Garden has continued to produce here.

We have nearly as many land-sharing agreements (and methods of compensation) as we do farm operations here. We pay Brian and Jane an annual sum for the disproportionate area that we (and all our sub-leasers) use. Chrystal and Ilya and Melanie pay for their leases with labour.

Robin welcomes me to raid her leek and celeriac patches for my family's use. My favourite method of lease-payment, however, comes from Rebecca. Once a year, she and her partner serve us dinner, and Rebecca gives us several jars of her wonderful preserves. In addition, she usually gives me some seed, depending on what has produced well for her.

When Robin was working on a research contract recently, she learned that agricultural lease rates vary widely even within the small area of the Saanich Peninsula. As of 2008, farmers were paying anything from nothing (but securing farm tax status) to $500 per acre, per year. There is no standard formula for land-lease arrangements. If you're planning to lease land to farm (or have land that you'd like to lease out to a farmer), work out whatever arrangement makes sense for you. Remember that the landowner likely paid a significant amount of money for the land, may be paying a mortgage, and is certainly paying property taxes. On the other hand, regardless of the market value of the land, organic farming is not a high-income activity. The rents that farmers pay are of necessity very low compared to the price of agricultural land in this area. Furthermore, the farmer is probably helping to reduce the landowner's property taxes, and is maintaining part of the property. Any arrangement can work, as long as all parties know what they're getting into, and as long as communication is clear and open.

Where to Farm

The first thing many would-be farmers have to figure out is where they will farm. In my case, I was able to skip this step. I had the great good fortune to have free land to farm for my first three years. After that time, Lamont and I decided to buy my grandmother's portion of the land that we now share with my aunt and uncle. Fortunately, she gave us a great deal, because otherwise we probably wouldn't have been able to afford to buy land in this area. Robin and Rachel will tell you about other land-use arrangements. When looking for land to farm, you should consider the climate, size, slope, sun exposure, water availability, drainage, proximity to markets, and fertility. Of course, in a perfect world you would find a piece of land that was perfect in all these ways, and you would be able to afford it. The unfortunate reality is that in the Victoria area, where the climate is great for year-round growing and there is a huge market for local organic vegetables, property prices are sky high.

Because of these prices, most new farmers end up in a leasing or land-sharing situation. All such arrangements should be entered into with caution and communication! We are very lucky to own our land, but we do not own it alone. We have had to work with Brian and Jane through misunderstandings and differing expectations regarding such things as the appearance of the land, landscaping budgets for shared areas, et cetera. If you are going to be sharing land in any way, it is very important to think carefully about what issues may come up, and get plans down on paper for how all parties will deal with them. All parties must be very clear about their expectations, and communicate those expectations well. If at all possible, find someone near you who has been in a similar situation and ask for their advice. Did they ever come to blows about water use? Did the person they were leasing from object to tiller noise on Sunday mornings? Did anyone expect the property to look like Butchart Gardens? Did "just give me some veggies when you have them" turn into "feed my entire extended family twelve months a year"? Did "long term" turn out to be only two years?

Don't get me wrong—many land-sharing or leasing situations work wonderfully well for many, many years. They can allow agricultural land to stay in agriculture while providing an opportunity for new farmers to make a living. These arrangements can reduce landowners' taxes while improving the quality of their water and soil, transforming fallow lands into beautiful, biologically diverse spaces, and creating or growing healthy communities. The odds of all these good things happening are greater if everyone enters the arrangement with their eyes, minds, hearts, and ears open, and with pens in hand to write down all agreements.

The south-facing slope of Northbrook Farm.

Getting Started

We felt rather overwhelmed as the time to move to the farm approached and we realized we needed a plan. Naturally, this English major turned to books. Let me tell you, it's not as easy as the books say (not even this one). However, the reading was very valuable. Books gave me ideas about how to begin. Much of what I read didn't really make sense until I was in the dirt, but once I was there, a lot of information came back to me, and I could make sense of what I was seeing.

My most valuable learning resource has been the farming community. Remember, everything you're doing has probably been done before by someone else. Every contact I made led to another. As I sit down to write this I'm going on a trip down memory lane, thinking of all the people who helped and taught me, many of whom have become good friends. There were really just two starting points.

The first was Mary Alice Johnson. I heard that she taught a course, Organic Farming as a Business, at a local college. This course was exactly what I needed. She co-taught the class with Tina Baynes. I signed up, and learned a lot from the actual course material, but perhaps more importantly, I met Tina, who farmed in my neighbourhood and owned a box-delivery business with Rebecca Jehn. They became not only friends but also our very first customers, as they bought our vegetables to put in their boxes. That first summer, Lamont apprenticed with Tina. It was only one day a week (because we were up to our eyeballs in work on our own farm), but he learned a tremendous amount from her. Tina also told us about other groups and resources in

the community. Through those contacts, we met other growers who invited us to their farms to learn from what they did.

The second one was that shortly after we moved here, I got talking to my new next-door neighbour. He mentioned that the regional government was funding a drip irrigation workshop to encourage water conservation. I hadn't even thought about irrigation! Here was something else to learn. So off I went to the workshop. The content was great, but the people I met there turned out to be an even greater source of information. My neighbour introduced me to a certified organic grower in the area who told me about the upcoming Islands Organic Producers Association (IOPA) annual general meeting. IOPA is our local organic certifying body. I went to find out how to get certified. Not only did I get my application and meet lots of other people in the community, I even ended up volunteering for a committee! This was a wonderful learning experience, and it provided more opportunities to meet more farmers who invited me to their farms to learn, and who shared their knowledge with me.

One of the things my new friends told me was where they got their feed, soil amendments (such as lime, kelp meal, rock phosphate, et cetera), and equipment. From them I learned that a good supplier is more than a source of materials. A knowledgeable supplier can save you a lot of money and time in the long run, even if their prices aren't the absolute lowest. For example, we checked out an irrigation store in Victoria that had great prices but staff who didn't know a thing! When we compared them with our nearby irrigation shop in Saanichton, we realized that although our local store was a bit more expensive on individual items, we'd probably save money (and grey hairs) overall, because they were knowledgeable and helpful. They could help us design an efficient system that would actually work, and we wouldn't have to keep making expensive changes year after year.

It's a tough balance, because for the first couple of years you won't be bringing in much income, so it's very difficult to find the money to pay for the things that you need. When you start shopping around, make sure you introduce yourself as a new farmer in the area, and set up accounts everywhere you shop. Most suppliers charge much less to their regular account customers than they do to walk-in customers.

From my experience, I would say that the most important things to keep in mind in getting started are to be open, to admit ignorance, to ask questions, and to volunteer. There might be an amazing community around you that could be an invaluable resource. Maybe all you need to do to tap into this resource is start introducing yourself and asking questions. There are opportunities out there for formal apprenticeships, through SOIL across Canada, CRAFT in Ontario, and ACORN in Atlantic Canada. In the USA, there is the National Sustainable Agriculture Information Service (formerly known as ATTRA). If you search "Organic Farming Apprenticeships," you'll find several programs. For short-term placements, try WWOOF (World Wide Opportunities on Organic Farms), Global Lifestyles, or HelpX. Or you can always call organic farmers around you and offer to come by and work with them for a day, a week, or a season. Tell them you want to learn and you want to help, and who knows what kinds of responses you will get? While you're doing all this, a bit of book-learning will come in handy too. See our suggested reading list for some of our favourites.

A word of caution: you will never know it all, and you may never feel "ready." After reading, apprenticing, and being involved in your community, at some point you'll have to make the leap. Don't wait until you think you know it all, or you will never start your farm. We sure didn't. When people asked about our farming background, my balding husband Lamont declared, "Crops? I don't even know how to grow hair!" We were blissfully ignorant of the challenges we would face.

Animal Problems

When we arrived on the property, one of our earliest challenges was staring us in the face in the form of a big fat bunny rabbit. I don't mean a wily little brown rabbit that looked like it belonged outdoors (although we had our share of those too). The ones that were really causing problems were the feral rabbits that had overrun the place. You know, the offspring of someone's cute Easter gift that they later released into "the wild" when the rabbit started eating, pooping, or procreating too much.

Those creatures were everywhere. They'd sun themselves on our driveway and would only move when a wheelbarrow tire came too close or when I'd pick up a tool to throw at them. Our first attempt to save the crops was exclusion. We put up a short fence around an area of the garden. The fence was black nylon netting, and we buried it six inches deep to keep the rabbits from digging underneath. So, instead, they chewed through it. Plan B was a little more gruesome. One day Lamont snapped. We approached a bed of lettuce to see a bunny snacking. Lamont instructed

In recent years, wet springs have led to big slug problems. We've tried all sorts of remedies, but, to Levi's delight, hand-picking is a daily reality in the spring.

me to circle around and herd it toward him while he selected a grapefruit-sized rock. By the time Lamont lobbed it, the rabbit was dashing at full speed, zipping this way and that. Somehow, the rock found its mark and came down squarely on the rabbit's head. We ate rabbit stew that night, and that meal was the beginning of the end of my decade of vegetarianism. Lamont did manage to kill one more rabbit that way, but we knew it wasn't a permanent solution to our large problem.

The next step was to adopt Fleece, our golden retriever. That first year Fleece was in heaven. She'd eat entire nests of baby bunnies and carry around the heads of full-grown ones the way other dogs carry old tennis balls. It wasn't pretty, but eventually a kind of balance was established. We still see rabbits, but they're no longer a major threat to our livelihood.

The deer have been another story. In the first couple of years, we put up with a bit of deer damage because it wasn't enough to justify the cost of fencing our whole twenty acres. However, as more and more of our region gets developed (reducing deer habitat), as our dog gets older and slower, and as we grow more food closer and closer to the woods beside us, the deer problem has increased greatly. So, the fencing had to begin. Eight thousand dollars and many hours of work later, the problem is not yet solved. They've jumped over, dug under, burst through weak corners and gone around the unfinished end of fences, so the work continues. In retrospect, we would have been much better off fencing before the deer got a taste, because now that they know what's here, they are persistent. Live and learn.

Our heirloom Lillie Mae's Little White Cucumbers have a cult following among our market customers and chefs.

Lillie Mae and the North Carolina Scene

Lamont and I have been very happy to discover that the feeling of community among organic farmers extends far beyond our little peninsula and our island. After our first year of growing, we returned to Lamont's home in Lincoln County, North Carolina, for a winter visit. While there, we sought out organic farms. Partly, we wanted to eat good, fresh food, but just as important, we wanted to check out the scene. What we found was a small but passionate group of very dedicated growers. Among them was Lillie Mae Boyles. Lillie Mae is quite a character, with her thick Southern accent, gruff exterior, and strong opinions (which she is more than happy to share). She farms organically, not because of any market advantage, and I'd wager that she's never thought about the "branding" of her farm or her produce. She just grows vegetables the best way she can, because it's the right thing to do. She welcomed us into her home, and even sent us away with some of the heirloom cucumber seeds that have been in her family for generations. This winter, we went back to see her again, to tell her that "Lillie Mae's Grandmother's Little White Cucumbers" have become one of our best-known and best-loved crops. Again, she welcomed us warmly, and we were pleased to see that not only has her farm grown, but that her community has too.

When we go to North Carolina, we also visit Nathalie and Cassie, pig and poultry farmers who have become dear friends (www.ggfarm.com). They have been instrumental in developing the market for local, organic products in their area. With Cassie's background as a chef and the pair's seemingly limitless energy, they've been able to connect with chefs and eaters who are passionate about food, and create a thriving food scene.

This is particularly encouraging to us, because when we first moved to southern Vancouver Island, we thought that the location, the market, and the food culture here were the main reason for the success of the local small farms. We would say to each other "What we're doing is great, but it's only a drop in the bucket of the food system. We could never make this work in Lincoln County." We thought that there were only small pockets of consumers in North America who cared more about the quality than the price of their food. The local food scene in Lincoln County is still a few years behind the scene here, but it's happening. If you grow good food and educate people about why it costs more, they will come.

What to Grow

After finding your dream piece of land, and while learning all you can from books and other farmers, you'll have to decide how much land to put into production. Farmers on small parcels simply farm all the land they possibly can. We had the wonderful and unusual situation of having more land than we could use for our first few years. We read, thought, asked others, and then ended up ploughing way too much! We had a great plan. We'd plant a quarter-acre in each of four crops: beans, beets, squash, and Swiss chard. In addition to that, we'd have an "experimental garden" in which we would plant small amounts of lots of different crops for our own consumption. We then planned to have another acre and a half in green manure (crops such as buckwheat or clover that are later tilled in to improve soil fertility). The following year we would put veggies where the green manure had been, and vice versa. Then, on top of all that, we had another two acres that we planned as a future orchard. We had a terrific plan to rotate several green manures on that piece over a period of two years, and then plant our orchard.

Our plan was solid. Our execution was not. Actually, the first problem was that the plan was overly ambitious. We hadn't taken into account the number of rocks in the soil, and therefore the amount of time that would have to go into picking them out. We didn't really plan any time for the green manures, and let me tell you, with only a walk-behind rototiller, prepping and planting large areas in green manure takes some time. We didn't get nearly the area planted in our four major crops that we thought we would. In retrospect, this was fortunate.

You see, we also had no idea just how much produce would grow in a small area. What we had intended as our experimental garden ended up providing lots of saleable crops. There is no way that the two of us (and all our family, friends, and acquaintances) could possibly have eaten all of those "small amounts." On the other hand, we would have gone insane with the tedium and backache of picking a quarter-acre each of beans and squash. We also learned that at the farmers' market, variety sells.

One problem with our ambitious plan was that in the first year, we opened up a lot more land than we could manage. We hired a neighbour with a tractor to plough and rototill what had been hay fields, so we wasted money by turning too much. Worse yet, once the land was ploughed, the weeds moved in. I think there was also a negative psychological effect. We started our farming career feeling as though we were out of control, and it took about five years for that feeling to abate. I contrast this with Rachel's experience. She started small and has gradually grown more and more each year. Sure, she feels too busy in the summer (there really *is* more work than you can possibly do at the height of the season), but she has always seemed to me to be so much calmer and more grounded than I am. And I know for sure her weeds have never been as out-of-control as mine. When I read her chapter, I don't get the sense of panic that I recall from the first couple of years that I farmed.

As a general guideline, I'd say that it's reasonable to start off with a quarter- to a half-acre per full-time person on the farm. When I wrote the first draft of this paragraph, I wrote, "With our degree of mechanization, I think that one acre

per person works, once you have a bit of experience." Since then, I've seen Robin and Andrew grow more food on one acre with three full-time people than I was growing on two acres with two full-time people. We're gradually learning that production is tied much more to labour than it is to area.

As you gain experience, you'll learn what suits your personality. Maybe you'll be an Andrew, able to coax stunning yields out of small areas with your attention to detail and your patience. Or maybe you'll be more like me—preferring to have more area under production so when you're a bit sloppy with the seeding or the watering, you can shrug and plant another bed.

We've all found that the labour crunch comes on harvest days. One person cannot harvest one acre of mixed vegetables that are producing well, and still have time to maintain them, so you get into hiring others to help. This may be a great way to go, but I'd suggest doing it on your own for a while first. In the meantime, if you're open to it, you'll probably have friends, family members, and even interested strangers offering occasional help, often just in exchange for vegetables. Since we live in the beautiful tourist mecca of southern Vancouver Island, we often hear from friends who want to visit. We love having guests, but we always tell them that only the first night is free; after that, they have to pitch in on the farm.

I can't tell you what you should grow, but here are some things to keep in mind while you're deciding. Grow what you like to eat. Not only will meal times recharge and inspire you, but crops are actually easier to tend and to pick if you're excited about how they'll taste. You will also be a much better salesperson if there is genuine enthusiasm behind your recommendation of the veggies. Talk to other growers and to buyers such as box-program owners, chefs, and produce managers. You can ask for general ideas of what they need, or you can seek contracts to grow specific crops. If you can find other growers in your area, it is great to run your ideas past them. Maybe the local produce manager is telling everyone that she wants more green beans and all the organic farms in the region have decided to plant more this year. In that case you might choose to grow something else or seek a new market. Also, during all these conversations, remember to talk about price. Maybe the reason no one is growing zucchini for the local restaurant is that they don't want to pay more than $0.60 a pound.

In choosing how many crops to grow, there are two things to balance. On one hand, you could grow only a few crops. This will keep your day-to-day operations and your recordkeeping simple while you're learning. You might be

Jarrahdale squash. One of the joys of the squash harvest is the sheer beauty of the different varieties.

able to learn a whole lot about growing a few things. However, you won't have the satisfaction of a bountiful market table or farm stand full of appealing variety. You also will miss out on the thrill of eating meals that are all yours. Perhaps more seriously, if one or two of those crops fail, you will have lost a significant portion of your income. On the other hand, you could grow small amounts of many crops. In this case, you'll learn a little bit about growing a lot of crops. You'll probably love the variety but curse the complexity. It is difficult to find the time to attend to the needs of many different crops at different stages of growth. I write lots more about recordkeeping later, but it is very important to keep thorough records, or you won't remember all the invaluable knowledge that you gain each year.

The first several years will be an experiment. You'll gradually learn what grows well for you, what you enjoy growing, and what makes money for you. Careful recordkeeping will tell you at a glance at the end of the season how much money each crop brought in. I aim for an average of $1.50 per square foot outdoors (that's $600 per bed, since my beds are four feet wide and a hundred feet long). Some labour-intensive crops, like salad greens and strawberries, do much better than that, and that offsets the lower-maintenance crops like winter squash and melons. If you have the space, these low-maintenance crops are great to cover ground. For me, winter squash and melons make around $0.50 per square foot. If your space is tight, you can focus on higher value crops. Also, in our climate you can plant in succession, so many of your beds can produce two crops each season. My average is fairly conservative, and accounts for some less-than-perfectly-tended beds. Robin pays much greater attention to detail and can expect $2.00 per square foot. Whatever your average, it's an important piece of information when you're deciding which crops to grow, and which to pass up each spring.

The colours and shine of our summer squash mix attract attention on our market tables.

Growing Squash—Heather

Summer squash is one of my mainstay crops. It's easy to grow, and produces prolifically. So prolifically, in fact, that by the end of August I start wishing for that first frost to wipe them out! I start my summer squash widely spaced in trays around the second week of April. This gives them time to grow their first true leaves in time to harden off, then be planted out around mid-May. Dates in your area may differ; just make sure you're planting them out after any danger of frost. The seedlings grow very fast, so they would probably like to be started in pots. I always run short of space in my greenhouse, so they have to make do planted twenty-four seeds per tray.

I transplant the seedlings two and a half feet apart in single rows six feet apart. Recently I've begun alternating squash rows with beds of other, smaller crops. This seems to allow better air circulation and thus slows the advance of powdery mildew and blossom-end rot in the late summer. Squash will survive under most conditions, but are heavy feeders that love lots of water. I work a layer of compost (approximately an inch thick) into the bed before transplanting, hand-water the plants in, then rely on drip irrigation. I think the best way to ensure good, early production is to cover the plants with floating row cover at transplanting time, and to leave it on until they start to flower.

The key to marketing summer squash seems to be variety in selection and size. Each year, the glut of medium-size to large green zucchini seems to hit barely a month into the season, and prices start plummeting. On the other hand, my bright, colourful mix of pattypans, crookneck squash, and several other unusual varieties sells well throughout the season, so I don't drop my price. I sell this mix as "baby" (pattypans roughly an inch to an inch and a half across) for $4.60 per pound, or "medium" (two to three inches) for $2.30 per pound. The few that we miss harvesting at this size are then sold larger, either at farmers' markets or at a price-sensitive store in our neighbourhood for $1.50 to $2.00 per pound.

My method of growing winter squash is similar to summer squash. I start these seeds in pots rather than trays, because I often leave them inside a bit longer before transplanting. This is because I don't feel rushed to try to have the first winter squash on the local market. However, it is important to give them a good start so they will have time to fully ripen in the field before the fall gets frosty and wet.

One more word of caution: beware niche markets. I know this runs contrary to conventional farming wisdom. It may be that you have a truly original idea, and that you'll be able to make a million and live out your farming dreams growing a niche crop. However, I think it's more likely that the niche market won't be a truly original idea. When we were still in the planning stages of our farm, Lamont and I went to an agriculture conference in Chilliwack, BC. There we met a man who told us, "I lost my shirt in cranberries, then ginseng, then kiwi. Now I'm a consultant." (Perhaps I should add another word of caution: beware of consultants!)

Around the time we moved to the Saanich Peninsula, many of our neighbours were digging up and burning kiwi vines. Several years before, word had gone out that our climate was perfect for kiwi-growing (it is), and that the price was right (it was); anyone could make a decent living growing kiwi. Well, not too long after, the bottom fell out of the kiwi market. Now many of the people who ripped out their kiwi vines are planting grapes. I'm scared of putting all my eggs in one basket. I think it's more sensible to plant a variety of crops. Fads come and go, but people have been eating lettuce, carrots, beans, et cetera, for a long time. Of course you want your produce to stand out, but to do that you can focus on growing unusual varieties, and on making sure that they are clean, sold immediately after picking, and of top quality.

How to Grow It

Field Layout

We had made lots of contacts, and some friends, in our new farming community. We had decided what to grow, and how much of it. Now we had to figure out how we'd grow it! The first thing to decide was how the field should be laid out. We were lucky enough to have a field on a beautiful, gentle, south-facing slope. The field is approximately four hundred by three hundred and fifty feet. We read that it was best for soil conservation to till across the hill, rather than up and down, so our first decision was made.

We then decided that we would plant in beds of several tightly spaced rows. This seemed to make sense on a farm that wasn't very mechanized. Tractors often plant in individual rows, but beds were better for us. Beds are a more efficient use of space than rows, because so much less area is lost to paths. It also means that less area gets compacted by feet, wheelbarrow tires, et cetera, and thus soil structure (tilth) and soil life are encouraged. If the beds are in the same places year after year, you can concentrate on amending the fertility just in the beds, not in the walking paths, so the money and time you spend building your soil is used to the greatest advantage.

We decided to make our beds four feet wide. This is slightly too wide to be convenient, because reaching all the way to the middle to harvest is a bit awkward. However, the wider the beds, the more efficient the use of space, and four-foot beds work perfectly for the tractor we ended up buying. The tractor tires straddle the beds and only compact the paths. Our paths are twenty-six inches wide, which is adequate for kneeling to weed and harvest, and leaves room for a bin for gathering crops.

We then had to decide how long to make the beds. At first, it seemed logical to run them from one edge of the field to the other. Up one side of the field is a gravel driveway, and the main water line, so we thought the beds (and irrigation lines) would start there and run all the way to the other edge. Somewhere along the way, Lamont realized that looking down a three-hundred-and-fifty-foot-long bed that needed weeding would be daunting indeed. Also, we learned that the type of irrigation we wanted to use would not emit evenly over such a long distance. The plants at the beginning of the bed would get more water than those at the end. So we decided to break up the field.

We made everything square (as much as possible): one hundred by one hundred-foot blocks divided into sixteen beds in each block. The field looked like a tic-tac-toe board with nine equal blocks (plus two smaller areas at the top where the field narrowed). Having a consistent bed size makes planning easy. A bed is a standard unit, so when we're comparing yields, for instance, we know that a bed in one part of the field is the same area as a bed in another part of the field. With all the beds and blocks being the same size, irrigation is interchangeable. This system also leaves main pathways (seeded in grasses and white clover) between the blocks that are wide enough to allow for driving in the field when it is dry. We've been very pleased with this layout. We now have a tractor that is a bit big for our needs, so the hundred-foot beds seem too small, but for all the hand work, and for the irrigation system, hundred-foot beds are ideal.

Irrigation

In our first years, we relied almost exclusively on emitter line or drip tape, a thin-walled hose is perforated to release water evenly along its length. This minimizes water use, and also helps with weed control, as the drip tape leaves the surface dry and thus does not encourage weeds to germinate. We chose line with emitters that drip water every twelve inches. In our clay soil, water easily spreads out horizontally, so having emitters only every foot worked well enough in the early years. After five years, our drip tape began to wear out, and as we gradually replace it, we are changing to drip tape with emitters every six or eight inches. We've noticed that as our soil structure improves with more organic matter, the water does not spread out horizontally as well as it used to, so the emitters spaced at twelve inches no longer water the whole bed as effectively. If you have sandy soil, you'll need your emitters closer together. Initially each of our beds had two lines of drip tape, and this remained the same whether we were planting four rows along a bed (as with lettuce), or two (as with summer squash). We now find that we're sometimes adding a third line to even out the moisture across the width of the bed.

Our system worked quite well on crops established early in the spring, but newly transplanted crops struggled in the heat of summer, with only drip irrigation to sustain them. We also discovered that in the dry season, nothing would germinate without moisture right on the surface. We later devised a better system. When first planting a bed, we lay a half-inch line right down the middle, into which we have inserted micro-spray emitters. Each one sprays out a two-foot radius, so the little spray just covers the bed without wasting too much water on the paths. Once the seeds have germinated, or the transplants have become well established, we replace the spray line with the two or three lines of drip tape. This is the system we

use on carrots, beets, lettuce, beans, parsley, basil, chard, et cetera. On salad greens, which like overhead water, we leave the spray emitters on for the duration of their production.

For some more widely spaced crops we have a different system. For melons and winter squash, which we plant two and a half feet apart, we custom-make irrigation lines. We use half-inch line into which we insert emitters that drip at a rate of one gallon per hour. We put one emitter every two and a half feet. Then, when a bed is all ready for transplanting, we lay out the irrigation, turn it on for a minute, then simply transplant into the wet spots and tuck the irrigation line right in beside the plant. This way, we're watering just the plants, not the weeds. Tomatoes only get drip irrigation (they don't start with the micro sprayers), because they don't like to have wet leaves.

Irrigation systems are surprisingly individual. What works for your neighbour might not be quite right for you. Some of our friends prefer overhead irrigation. One of our friends, who farms just fifty kilometres away, finds that seeds germinate just fine under her drip tape, but she prefers a different brand from what we use. She finds that the brand we use moves around on the bed too much as the tapes expand and contract with the differences in day and night temperature.

Crop Rotation

Another consideration when you're laying out your field for the first time is your long-term crop rotation. Check our recommended reading list for sources that are good for crop rotation and fertility information. I especially like Eliot Coleman's *New Organic Grower* for clear crop-

Pole beans require more work than bush beans at planting time, but provide a longer harvest and are easier to pick.

rotation plans. The ideal rotation plans I had in my first couple of years have changed a bit as I have learned what does and does not grow well for me. A rotation plan works best if you want to grow exactly the same amounts of many different families of plants, and if all areas of your land grow all those things equally well. The reality for me is that some crops grow very well (all cucurbits, the cucumber, squash, and melon family) and some crops grow poorly (most brassicas, the broccoli and cabbage family). Furthermore, not all areas of the field grow all crops equally well. For instance, every year I'd love to have the high, drier areas of the field in early spring crops followed by the overwintering crops that I plant in late summer. I don't do this, however, because the soil would never get properly rested under a cover crop, and because too many of the spring and winter crops are brassicas. There wouldn't be enough variety for an adequate rotation.

Recently transplanted tomato plants in Heather's Tomato Tunnel.

Tomato Tunnel—Heather

For several years, my tomato dreams have exceeded my greenhouse space, so I have grown some outdoors. Some years I have had moderate success with this, although I never had the yields that I could expect under cover, and some years fall rains bring the late blight so early that I get very little yield. A few years ago, I heard of a solution much cheaper than a permanent greenhouse: a simple, inexpensive hoop house. Four-foot lengths of rebar form the footings for the PVC ribs of the structure. These pieces of rebar are pounded in two feet deep, every four feet along each side of the hoop house. The width of the structure is ten feet. In my case this perfectly covers two beds and the twenty-four-inch path between them. The ribs are twenty-inch PVC

pipe, bent so the ends slip over opposite pieces of rebar. It's best to use three-quarter-inch pipe, but you can skimp a bit and get away with half-inch if you're not in a very windy location.

At this point, pause to figure out how you're going to support your tomatoes. I use twine, wound around the main stem of the tomato plants and tied to wires that are strung tight between two rebar stakes. These stakes are pounded in to give stability. Put in whatever trellising or supports you'll need now, before you put the plastic on the hoop house. I use a hundred- by twenty-foot vapour barrier, and I make the tunnel a hundred feet long, so one whole roll covers the structure perfectly. Pull it over on a calm day. Next, I cover the whole thing in twine in a criss-cross pattern. The twine is all that holds the plastic on. I tie the twine onto the rebar at ground level (just below the pipe) in one corner, then cross it over the tunnel, skip one rib, and tie down to the rebar under the third rib, and so on down the whole tunnel. I then start back in the opposite corner and repeat the process. To keep the end ribs from bending inwards, I tie them to big, well-pounded T-posts at each end, and secure the plastic around the ends with lots of Tuck Tape (red tape made for sticking on clear poly). I don't put any cover over the ends, but if you want to, it would not be difficult to rig up a curtain-type arrangement. In the fall, when the heavy winds begin, we simply take it all apart, roll it up, and put it away for the next spring. One of the benefits of this system is that I can move my tomatoes each year.

One year I experimented with something new. I set the tunnel up in the cold, early spring and seeded a bed of beets and a bed of carrots inside. I thought they would germinate early, giving me a head start on the season. When the tomatoes were ready to transplant out, I took down the tunnel and reassembled it where I wanted the tomatoes. Unfortunately, the beets and carrots were a failure. I think several things went wrong. I was foolish to think that this structure, without end walls, would trap any warmth. The cool winds blew through, and the soil didn't warm enough for the seeds to germinate. Some of the seed did eventually germinate, but I had planted them in poor soil that I had worked too early, while it was too wet. This damaged the soil structure, impeding the growth of the already-stressed little plants. If I try this plan again, I'll wait just a little longer so the soil isn't so wet, and then I'll put end walls on the tunnel.

As a general guideline, I try to balance these many factors by making sure that each year I set some areas aside to be in green manures. A green manure is a crop grown to improve soil structure and fertility; it is turned in rather than harvested. I'll explain this in more detail in my section on soil fertility. I have also now incorporated chickens in a movable pen (more on that later), so there is always some part of the field that is getting a rest from annual vegetable crops. During its rest, the fertility is being restored, either with green manure or chicken manure, and cycles of disease and pests are disrupted. In addition, I keep careful track of where each crop was planted each year, then I just move things around, being especially careful not to plant the same families in the same locations two years in a row. Sometimes this means that a crop is not in its optimum growing location, but I feel the benefit to the soil of a thorough rotation outweighs the inconvenience of not being able to put everything exactly where I'd like to.

Fertility

When I started out, I was very intimidated by the reading I did about soil health and fertility. All this talk of N-P-K (the macro-nutrients Nitrogen, Phosphorous, and Potassium)—how would I know what I needed and how much? Sure, I understood that my compost should have a good carbon-to-nitrogen ratio, but how would I know if I achieved it, and then how would I know how much compost to apply? Then I came to the sad realization that compost doesn't grow on trees. To apply it at the rates recommended by some books (especially the organic home-gardening books), I would have to spend all my money having materials trucked in (manure, hay, seaweed) and all my time turning and hauling the stuff. Then of course there were all the micronutrients to consider! I thought a soil test would be a good idea, but then I heard that these could be inaccurate, or that the results could vary with the time of year, the temperature of the soil, the point in the growing cycle when the tests were taken, et cetera. I heard of an old, experienced farmer who said he took a soil sample, divided it in half, sent it to two labs, and got two different results, and I have heard a soil scientist say that testing for nitrogen is imperfect at best.

Although I had reservations about soil testing, I had to start somewhere, so I sent a sample off to a lab. Perhaps the most valuable piece of information from the test was my soil pH. We live in an area of naturally acidic soils, and most vegetable crops like relatively neutral soil, so the application of lime is an ongoing necessity here. Beyond that, I learned that my organic matter was low. Now, with a bit of experience, I could have determined that just by looking and feeling the soil. I also got recommendations for the amount of nitrogen, potassium, and phosphorus to add per acre. I divided this out to tell me how much should be added to each bed, then started buying amendments. That first year I used a lot of alfalfa meal for nitrogen, because I hadn't been on the land long enough to make compost or grow cover crops, also known as green manures. In addition, I added rock phosphate and sulphate of potash to each bed, along with a handful of kelp meal for micronutrients. Since that first year, I test only occasionally, when I see poor crop performance that I cannot figure out on my own.

I have come to believe that in general, soil life is more important than specific N-P-K numbers.

Soil life depends on organic matter, on cultural practices (mulching, tilling, crop rotation, watering, et cetera), and of course on nutrient levels. I highly recommend the book *The Soul of Soil* for a good background in basic soil science, and then I think it's most important to watch your crops to find out what they need. Deficiencies show themselves in the plants. Good quality compost not only adds nutrients, but, perhaps more importantly, adds beneficial microbes to the soil. But compost is not the be all and the end all of healthy soil.

Building compost piles can be useful, and indeed some farmers manage their compost very carefully. However, a lot of "composting" can happen right in the soil when you incorporate organic matter. This could be crop residues, green manures, animal manures, or other plant material brought in from off-farm. This saves time and energy compared to moving material into and then out of piles, and can save nutrients that might be lost through leaching or dissipating in the air when compost piles get too wet or too dry, or are turned. Keep your eyes and ears open for sources of organic matter in your community. For example, in Victoria we have an organic tofu plant, and the owner donates the okara (what is left of the soybeans after they make the tofu) to local organic farmers. This is a great source of organic matter, and is fairly rich in nitrogen. Naturally, it's also in very high demand and there is never enough to go around. There may be restaurant owners near you who are more than happy to save their kitchen scraps if you pick them up on a regular schedule. These can make great additions to a compost pile. Beaches can be a source of seaweed. Neighbours with animals might give you manure. If you're planning to be

Laying hens forage in a overwintered cover crop that was left for chicken pasture rather than turned in in the spring.

certified organic, however, remember to check the status of everything you plan to bring onto the farm, to make sure you can use it, then document it when you bring it onto the farm.

Chickens

Several years ago we added chickens to our farm. This idea intimidated me at first, and we probably wouldn't have done it if it weren't for a friend who was getting out of the business and offered us her slightly used laying hens. I didn't feel like much of a farmer at the beginning of my foray into animal husbandry. When the appointed day arrived for me to pick up the chickens, my friend wasn't going to be there, but she said she would leave them in their coop so they would be easy to catch. Well, it turns out that there's no such thing as an "easy-to-catch" chicken for a gal who had never owned an animal bigger than a newt. Not only was I hopeless at catching them, but much to my embarrassment I discovered that I was afraid of them. Yup, I was intimidated by your basic domestic chicken. Fortunately, a friend had volunteered to help me. He ended up catching all the hens.

The flock I acquired was all female, and since our regional organic standards require a rooster for normal socialization of poultry, I mentioned to my friends that I was on the lookout for a rooster. Very soon, I got a call from a friend who worked in a nearby industrial area. She had heard of "C.D." (short for Cock-A-Doodle), the stray rooster. It seems he showed up outside the door of a cafeteria one day and never left. The poor creature was harassed by dogs, had no other chickens to keep him company, and subsisted on what he could scrounge, along with leftovers provided by the concerned cooks. These cooks were delighted to hear that C.D. could live out the rest of his days on my organic farm. Again, it sounded easy enough: "Just c'mon by and get him." This time the friend who offered to assist had no more experience than I did. Fortunately, I had handled my hens enough to be getting over my fear, and C.D. was a bantam (a small breed) so he did not look intimidating at

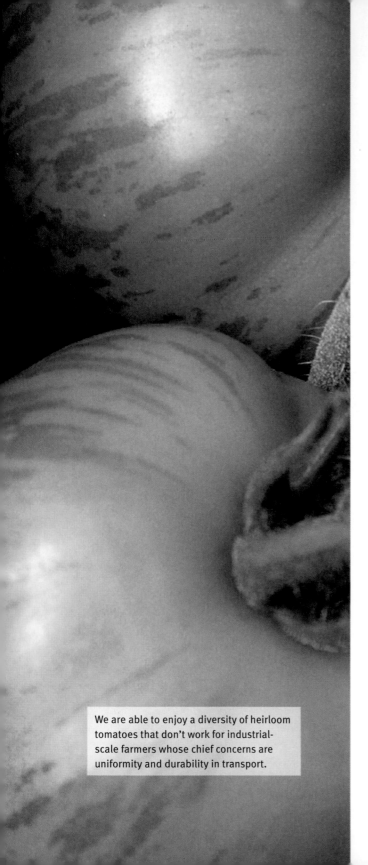

all. However, what he lacked in size he made up for in speed and agility! My hens would come running when they heard the feed hit the pan, so I assumed I would be able to lure C.D. with food. No such luck. After half an hour of running and diving after a flapping rooster, we felt both incompetent and foolish. How the heck were we going to get that bird into our cage? Just as we were about to give up, the nearest door opened.

A tall, dark, handsome stranger emerged. "You ladies need some help?" he asked.

"Are you any good at catching chickens?" I laughed.

With a shrug and a barely perceptible nod, he calmly strode up to C.D. and grabbed him on his first attempt. Off he sauntered into the sunset, and we never even learned the name of our rooster-rustling knight.

When we first brought our flock home, we tried "chicken tractors"—small, movable pens that I could pull by hand onto fresh ground every day. This might work beautifully in some situations, but I didn't like it. I wanted to leave the chickens in one place long enough for them to clean it out, and then wanted to plant a cover crop before another food crop (to adhere to the organic standard, which prohibits growing a food crop after the application of raw manure). I found it awkward to have such small patches at different stages of growing and grazing, so we quickly changed our plan.

Here's what we do now: we house the chickens in a coop that can be dragged around behind the tractor, and enclose this in a fence that gives them a run of approximately a hundred by sixteen feet. In about two or three months, our thirty chickens can totally clear the vegetation out of an area this size, and then it's time to move them. The old chicken run gets planted in a green manure crop; then, when it is turned in, we have a lovely, fertile, rested area ready for a heavy-feeding vegetable crop. I find that the areas where the chickens have grazed grow great vegetables. The soil is fertile and diseases are much reduced. I think this is due to the increased biological health of the soil. Weed pressure, however, is not reduced. The chickens don't seem to dig out weed seeds, and their food sometimes contains viable seed that gets spilled as they scratch around. Oh well, if the chickens took care of everything, it would be too easy!

I hate infrastructure in general, fencing in particular, and I have a natural tendency to take the easiest (or laziest) route. One year, at the end of a bad blueberry season, I had a brilliant idea for curing my mummy berry problem. Mummy berry is a devastating fungal disease that prevents the berries from ripening properly. The fungus spreads from fallen, infected berries. I put my chickens, their coop, and their feed in the middle of my two-acre blueberry field, hypothesizing that they wouldn't venture far. I didn't bother putting up a fence. I had visions of them eating every fallen, diseased berry, thus curing my mummy berry problem. At the same time, they'd eat all the weeds springing up through the sawdust mulch, and deposit their rich manure around the plants. Well, I discovered how far the chickens will venture when unfenced: as far as the nearest compost pile. Every morning they'd beeline from the coop, all the way through the field to the pile, where they'd scratch, dust bathe, and feast on my lovely red wiggler worms. The few who tended to linger around the coop began to get picked off by birds of prey, so, it was back to the fenced area, complete with netting over the top of their run.

Melanie demonstrates the Whizbang Chicken Plucker, built by our friend Cam from a design he found on the Internet. It saves us so much time by plucking chickens in about 15 seconds when it used to take us about 10 minutes by hand.

Whizbang Chicken Plucker

A farmer from the US, Herrick Kimball, published a set of instructions for making a low-cost drum plucker for chickens, and sparked a revolution for backyard chicken processors. (See YouTube videos—not for the faint of heart!) Plucking takes under a minute now, whereas before it was an onerous chore. Although building the plucker turned out to be both more difficult and more expensive than first described, it nonetheless has improved our lives. A friend of ours ordered the plans and built the machine. Four local farmers chipped in to cover the costs, and two more have since bought in because it works so well. The plucker is on wheels, so we move it from farm to farm when birds are ready for slaughter. http://whizbangplucker.blogspot.com

When the hens' egg production drops to the point that I cannot sell enough eggs to pay for their feed (after roughly two and a half years), I slaughter them. They are skinny and tough compared to chickens grown for meat, but they provide our family and friends with wonderful, flavourful stews and soups. In recent years I have begun raising a flock of meat birds each summer. These are the quick-growing varieties of chickens selected for their meat. They live only ten weeks or so, and during that time I keep them in a chicken tractor. I then slaughter them on the farm, and pluck them in the Whizbang Chicken Plucker that our dear friend Cam built.

My sons are aware of the practical role of animals on the farm, and they help out in little ways on slaughter days. I didn't realize how pragmatic they had become until I mentioned, within five-year-old Jackson's hearing, that I thought we should get a new dog.

"Why?" he asked.

"Fleece is getting too old to chase away racoons and deer," I replied.

"Will we slaughter her?" was his innocent question.

I learned a bit myself about the seemingly arbitrary lines we draw between species when I explained to him the difference between farm animals and pets.

Green Manures (or Cover Crops)

I'm a huge fan of green manures. By planting these crops, you are using the miracle of photosynthesis to harvest energy from the sun and store it in your soil. Green manures give you many of the benefits of composting in a very efficient way.

Instead of having to haul compost around by tractor or wheelbarrow, it can grow right where you want it.

Typical green manures include buckwheat, grain and legume blends, and clovers. All are beneficial in different circumstances. Buckwheat is a great choice for spring planting, since it grows quickly and tolerates drought. It is easy to turn in and decomposes quickly, so food crops can be planted shortly after a buckwheat crop. Although buckwheat doesn't fix nitrogen from the air, it preserves the nitrogen in the soil by preventing its loss though leaching or dissipating in the air. The nutrients that the buckwheat holds then become available when it is turned into the soil. Buckwheat also prevents weeds from taking over fallow areas. It adds a lot of biomass to the soil as it decomposes, and until then, its lovely flowers provide great forage for bees.

Over the winter, a fall-planted mixture of 80 per cent rye or winter wheat and 20 per cent peas and/or vetch makes a great cover crop. The grain adds organic matter and a good root system to prevent erosion, while the peas fix nitrogen from the air into nodules on their roots, which is then released when the crop is turned in. When I was first learning about the wonders of legumes

Before I started farming I had no idea that there are scores of different variety of carrots, each with unique taste, appearance, or growing requirements.

for adding nitrogen to soil, I wondered why we shouldn't plant just peas and get even more nitrogen! For one thing, the grain acts as a pole for the legume to climb, but more importantly, legumes fix nitrogen only if it is not readily available in the soil. This is where the grain comes in: the sprouting grain takes up the available soil nitrogen, forcing the legume to fix it from the air. Eventually, the nitrogen that the grain took up is also returned to the soil when it is turned in.

With green manure crops, you should turn them in just as they are beginning to flower, or just about to set seed in the case of grains. This is when the biomass is at its peak. If you let a green manure go to seed, it will become a weed problem. The crops turn in best, and decompose fastest, if you mow them first. Clovers work well as nitrogen fixers too, but they are harder to kill, and can become persistent weeds if you are not careful. However, we do like one type of clover for one situation.

In our first few years we were surprised at how much work growing green manures can be. We often had large areas that we wanted to improve with green manures for a whole twelve- to eighteen-month cycle. This meant a lot of energy went into planting, then turning in,

a winter crop, followed by one or two summer buckwheat crops, then another winter crop before returning to vegetables the next spring. We then learned about yellow sweet clover, which is a biennial (that is, if you plant it in the spring, it doesn't flower until the following spring). So this is a green manure that can be planted in the early spring, and left undisturbed for a whole year. Unfortunately, when we tried it, the clover was slow to establish, so weeds had a chance to get ahead. We had to mow several times to prevent the weeds from going to seed, but it was great not having to turn it in and replant with the changing seasons, especially since the less the soil is disturbed, the healthier it is. The clover was difficult to kill, and took quite a while to decompose, but the subsequent vegetables grew well.

All soil-building techniques require inputs: energy, equipment, and time—or some combination of these factors. For instance, large-scale green manuring can be difficult without a tractor. Rachel's section describes some pros and cons of mulching. Livestock requires infrastructure and care three hundred and sixty-five days a year. Whatever plans you have for building soil health and fertility will probably evolve (and perhaps change radically) as the years go by.

Lamont uses our spader to incorporate a beautiful green manure crop at Three Oaks Farm. Lush growth, especially in moist soil, requires two passes of the spader.

Equipment

One of the most difficult and intimidating things for me to wrap my head (and my cheque book) around in the beginning was equipment needs. In our first year we bought lots of hand tools (used), wheelbarrows, a good walk-behind tiller, a utility trailer, a sickle-bar mower, and a lot of irrigation equipment. There was a lot more equipment that we thought we could use, but we just couldn't spend too much money before we had any income from the farm. The next year we added a good Weed Eater (with a shoulder harness—essential!), a ride-on mower, and a tractor, but more on that later.

In our first year we hired a neighbour with a tractor to plough and rototill for us. For the next couple of years we still had that neighbour come back to till large areas. We certainly used our little tiller all the time, but for turning in green manures, or in the spring and fall when time is short and there are large areas that need tilling, the walk-behind tiller doesn't quite do it. That being said, we were not very pleased with the results of tractor-powered tilling. The big tillers turn the soil so fast and so thoroughly that they can create several problems. They damage the tilth or structure of the soil, and kill earthworms and other soil creatures, and they break up the soil so finely that it then becomes subject to compaction. Basically, they dramatically alter the soil ecosystem. In clay soil like ours, big tillers can also create hardpan at the depth where the tines reach. Hardpan is the layer of compacted soil below the loosened upper layer. It can be impermeable to water, plant roots, and critters. This can also happen with walk-behind tillers if you either till too much or till when the ground is too wet, but hardpan develops much sooner with tractor-powered tillers. For all these reasons, we gradually moved away from tractor tilling.

Instead, in our second year we bought a tractor ourselves. We got a used cultivator (imagine a heavy, steel set of curved fingers that drag through the soil), and the next year added a set of discs (picture a set of twelve angled pizza cutters that slice and toss the top four inches of soil as they roll behind the tractor). The cultivator is great for loosening the soil down to twelve inches deep and preventing hardpan. We used the discs for turning in green manure crops and residue from vegetable crops. After the discs and the cultivator, the soil was ready for our walk-behind tiller to do the final bed preparation. Unfortunately, we found we were going over the field many times to prepare for planting, thus causing significant soil compaction and still requiring the walk-behind tiller.

There is no easy answer to what implements you might need, and there are ways you can get creative. One year, out of desperation to save time planting green manures, I discovered that several passes with the cultivator, followed by one pass dragging a rolled-up chain-link fence to break up clods, prepared the ground enough to broadcast cover crop seeds. Another pass dragging the fence covered the seed. It certainly didn't do as pretty a job as a tiller followed by a precision seeder, but

it was good enough for me, and the cultivator is much easier on the soil than the tiller.

In our sixth year we made a big investment and bought a spader for our tractor. . Instead of the curved, rotating tines of a tiller, spaders have flat paddles that dig into the ground and push the soil backwards. This serves the same purpose as a tractor-powered tiller, but better, deeper, and in a more gentle way that does not destroy the tilth of the soil. Furthermore, the "spades" are placed so that they reach to differing depths in the soil. This means that no hardpan is created below the spaded area, a great advantage over tilling. The spader does not leave the fine powder that a tiller does, so we had to get used to seeding in coarser soil. We were a bit skeptical at first, but so far it is working very well. The spader is amazingly good at turning in crop residues. I can spade in huge squash plants, wait a week, spade again, and the ground is ready for the next crop.

However, sometimes one solution can lead to another challenge. Our spader was not a perfect match for our tractor. The tractor had barely enough horsepower to drive the spader in our heavy soil, and it couldn't go forward slowly enough for the spader to work at its best. So, two years later we bought a new tractor. The preceding season had been our most productive yet, but we reinvested all our profits. We saw a big difference in our soil when we started using the spader, and now we're seeing more improvement with the new tractor. This improvement is contributing to higher production, but the benefit is slow and more long term compared to the big financial investment in the equipment.

The decision to buy a tractor is a big one, and you may choose never to buy one. Tractors are so expensive that for that amount of money, you could hire others to do custom tractor work for many years. Repairs on tractors are costly too. However, if you have such a large area that smaller equipment is inefficient, you might have to go for it. If you can find a good used tractor with a front-end loader, you'll always find little jobs for it. The front-end loader is invaluable for turning and moving large quantities of compost, and this capability will help reduce dependence on expensive purchased compost and other soil amendments. Another big advantage of owning a tractor over hiring one is flexibility. You can turn in little bits of your field throughout the season rather than waiting until the whole field is ready to be tilled. This helps you stay on top of your green manures and your crop succession.

In general and when possible, delay buying equipment until you're pretty settled into what you're doing. That way you might avoid some costly buying and selling if you change your mind about how your field should be laid out, or even which crops you will grow. Try renting, borrowing, or sharing for a little while until you know exactly what you need; then you'll be more confident when you buy. One way or another, you'll need tools of some kind to help you mow, turn in cover crops, prepare seedbeds, plant seeds (or seedlings), weed, and water. Of course, these tools can be anything from scythes, spades, forks, hoes, trowels, and watering cans, to massive tractors with every implement you can dream of. The low-tech options have the obvious advantage of costing less, and you'll stay in good shape! Over your first couple of years farming, however, keep your eyes and mind open for slightly-higher-tech options that might save you time and backaches.

Our Favourite Hand Tools

Heather: If I had to live the rest of my life with only four hand tools (heaven forbid), I would choose one sturdy, long-handled digging fork (four tines), one long-handled spade, one hoe (probably the Lee Valley Slim Draw Hoe), and the Ho-Mi Digger. I choose the Ho-Mi, a traditional Korean tool, because it can double as a transplanting tool and a hand weeder. This four-tool scenario is a nightmare for me, though, because I'm a bit of a tool slut: I'm a sucker for a shiny new hoe in a shape I haven't tried. This isn't all bad, because tool preference is very personal; my apprentices and farmhands all have different favourites. It's worth spending a few hours volunteering on other farms just to have the chance to try all their hand tools before you stock up for your own farm.

Robin: Someone once said, "Robin evolved before the tool users," because I tend to use my hands a lot. I'll often weed by hand before grabbing a hoe but when I do, I prefer the half-moon hoe. It's good for making seed furrows as well as for weeding. While spades are indispensable in the garden, I love my square flat shovel for emptying soil out of the back of the

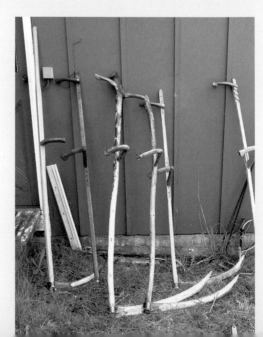

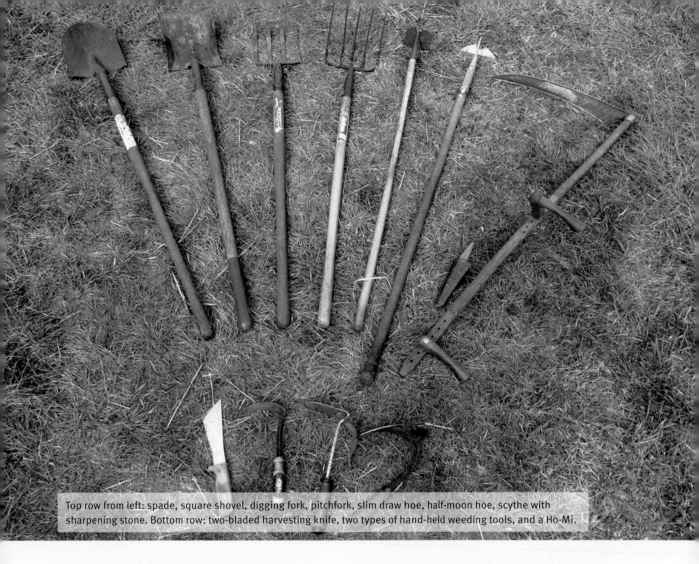

Top row from left: spade, square shovel, digging fork, pitchfork, slim draw hoe, half-moon hoe, scythe with sharpening stone. Bottom row: two-bladed harvesting knife, two types of hand-held weeding tools, and a Ho-Mi.

truck or trailer. It glides along the floor of the truck and leaves a perfect clean swath behind it. I use a digging fork for loosening soil around tap-rooted weeds and for some harvesting jobs.

Rachel: My top hand tools would be a spade, a pitchfork, a slim draw hoe, a Ho-Mi, and a scythe. I choose tools that are as light as possible without compromising quality. I prefer a five-tined pitchfork to a digging fork because it covers more area, and the rounded, sharp tines slide into the earth easily. The curved tines save you energy. The slim draw hoe from Lee Valley Tools is lightweight and has

a long handle, and the blade is dense enough to handle some heavier weeds. The scythe is my new favourite tool. It is indispensable for cutting back overgrown beds to prepare for the next planting. Cutting back, raking off, and then preparing to till is far faster than hand-pulling the plants, and is more satisfactory than the Weed Eater. I use my scythe to cut grass which is then used for mulch on my asparagus and fruit trees.

During the years when Heather did the Saanich Organics administration, this was a common sight: a sales call to a chef while continuing with other farm work.

The Harvest!

Once you have worked like a dog all winter, planning, then all spring, preparing soil, planting, watering, and weeding, the day will come when something grows. At this point I remember thinking, "Holy crap! Now I have to figure out how to harvest and sell this!" Harvesting is an area where you can learn a tonne by helping others, and by asking questions. I don't know many farmers who would turn down an offer of help on a harvest day, and the more farmers you can see in action, the more ideas for efficiency you will pick up.

Our general system is something like this: At the start of a harvest day, I make sure I have lots of plastic bins rinsed out and ready. Sometimes I even spread them around the field so they're ready at the end of each bed I will be harvesting from that day. Then the picking begins. I have elastic bands in my pockets, because I find it fastest to bunch chard, parsley, kale, beets, and carrots right there in the field. I have a little pen-style Newton scale I use in the field to recalibrate myself occasionally and to make sure my bunches are a consistent size. I carry the produce down to my packing shed, wash and box it, and at the end of the day, drive it all up to the garage in my truck.

During every step of the harvesting and packing process, I keep my eyes open for imperfections—any insect-damaged or wilted produce is culled. Sometimes this is hard to do. If you've put hours and hours into a bed of carrots that looked perfect until you sprayed off the dirt, it can be tempting to think, "Well, gee, that wireworm damage really isn't too bad," and sell them anyway. This is especially the case if you have already promised the crop to a buyer. In the long run, though, it's better for your business to eat those carrots yourself, or give them to family, friends, soup kitchens, dogs, or chickens, and only sell the really great produce. Sometimes this will necessitate awkward phone calls to chefs, to apologize that they won't be getting all of their order, but it helps to build up a reputation for top-quality produce that will allow you to charge the prices you need to make a living.

Our first harvest day was incredibly discouraging. It took Lamont and me all morning and half the afternoon to pick $47.60 worth of Swiss chard. We figured we had each made about $4.00 per hour for the harvest, and that meant $0.00 per hour for all the work that went into it before the harvest! Looking back, we had made several mistakes. We were picking the crop too soon. The leaves weren't big enough to make for efficient picking, and we had overestimated how much was in the bed so we were really scrounging to meet the order. We were washing in small batches in tubs on the ground, which was slow and hurt our backs. We just didn't have a system.

Now the situation is much improved. We have made all kinds of changes, including moving the packing shed to the bottom of the field, so we can walk downhill with the heavy, full bins, and uphill with the light, empty ones. It is essential to have a good stock of markers and masking tape for labelling bins, and a well-organized harvest list, now that our harvest is substantial. Disorganization eats up so much time!

When you are developing your harvesting system, remember to observe others and to experiment. There is no single right way to do it, but every tiny little thing you can do to improve efficiency and to reduce disorganization will really pay off over the length of a season. Although it's a good idea to try different methods, make sure they never compromise the quality of the end product. One key to our harvesting that has not changed from our first year is that we sell our crops before we pick them. We tell our customers what we have available, they tell us what they want, and then we pick exactly what they have ordered and get it to them as soon as possible. This way, they get produce that is truly fresh, and we don't have any waste.

John in our washing area. Adding a cement pad was a great improvement over the bare ground we used to have.

The occasional casual, outdoor potluck dinner during the farming season provides a chance to build community while enjoying the fruits of our labour.

Cooking

We find it ironic that new houses are being built with bigger and bigger kitchens, loaded with stainless steel appliances and gadgets of all descriptions, and yet it seems that relatively few people cook meals from scratch. Quality time can be spent together in the kitchen rather than watching TV. The food industry has done an amazing job of convincing us that cooking is an inconvenience, and that inconvenience must be avoided at all costs, but we don't want to grow huge quantities of a few crops to sell to processors. We want to sell delicious, healthy, simple food that will make it onto a plate by the most direct path possible. We feel it is part of our role to help reconnect people with vegetables. Not only do we sell them, but we also provide cooking tips and recipes. It is shocking to us how often we meet people who have no idea how to cook whole foods. It is very satisfying when they come back to the market the next week eager to tell us about eating their first roasted beet or steamed chard. We self-published a compilation of simple recipes from our box-program newsletters, which we sell at the market and to box customers.

If you'd like a copy of our cookbook, e-mail us at info@saanichorganics.com

There are lots of different marketing options out there, including farmers' markets, farm stands, restaurants, grocery stores, Community Supported Agriculture (or other box-delivery businesses), wholesalers, and probably others I've never thought of. When we started out, we sold mostly to box programs and the farmers' market, with a bit to restaurants and one small grocery store. Now all our sales are to Saanich Organics, the distribution company we own. We describe how it works in Chapter Five.

While you're establishing your markets, try to balance efficiency, price, and personal satisfaction. For example, it may be very rewarding to know each customer personally and it may be your dream to deliver custom-ordered boxes of perfect produce to families, but it is unlikely that you will have time to grow and sell enough to make a living that way. At the other extreme, for maximum efficiency, you could grow all one crop on a large scale and sell it through a distributor. But if you go that route, not only would you be unable to maintain the biodiversity and crop rotation necessary for a healthy ecosystem, but you might not be able to get the price you need. You'd be competing with the economies of scale (and subsidies) of huge farms all over the world. This model also may not be as personally satisfying for you.

Think about why you're farming, and that will give you clues as to how you should approach marketing. Is it a passion for growing? For food? For business? For people? This might help you decide if you want to maximize your time alone in the field or if you want more interaction with the public through a farmers' market or a farm stand. Weigh all these factors and make a plan, but be open to the idea that it may have to change year to year.

Wherever you sell, you will have to decide on a price for each crop. In the Victoria area, many consumers recognize that it costs more to grow organic produce here than it does in other parts of the world, so they are willing to pay more for local produce than they do for imported. This willingness has come about through the work of many farmers who have decided to keep their prices high enough to make a living, and then have put in the time to educate their customers about why they charge what they do. Because the farmers in this area try to be co-operative rather than competitive, we don't undercut each other's prices. This is extremely important. Wherever you are growing, try your best to develop this kind of community among farmers in your area. As long as people are eating imported organic vegetables, we should not be competing with other local growers for market share. Instead, we should be co-operating, so that we can all grow more food and enable more people to eat locally. No farmer benefits if we're all busting our butts just to sell vegetables cheaper and cheaper as we compete with each other.

Recordkeeping

Although every farmer I met as I was starting up told me to write down information right away, I still sometimes think, "Oh, I'm busy right now. I'll record this later." Well, believe me, no matter how memorable a detail seems at the time, if you don't record it, you will forget it. This becomes more true the longer I farm, as the years start to run together in my memory.

Why is it so important to be able to refer back to everything you do? So you can learn. Every seed you put in the ground is an experiment, and each of these experiments has countless variables (variety, location, date, weather, soil temperature, moisture, weed pressure, preceding crops, et cetera). The more of these variables you can keep track of, the sooner trends of successes and failures will reveal themselves, and the more you will learn from both. There are countless ways to keep records, and an unlimited number of details you could record, so you will have to draw the line somewhere. What follows is an outline of the elements of my recordkeeping system.

Drip tape provides the irrigation for our strawberries.

Farm Journal

I used to maintain a daily journal in a three-ring binder with a page for every day of the year. This was an informal place where I could write down anything that happened on the farm that day. I used a binder so I could add more sheets. I usually wrote only a few lines each day, and then would move on to the next page the next day. Then, the following year, I would go back and write on the page again. So, for example, July 1st was on the same page every year, and when that page finally filled up, I simply added a new page. Keeping the information from same date of different years all together made it easy to compare farm activities that were happening at any given time of the year.

Now the farm journal has simplified: we use one page for a whole week, rather than for a single day. Each week at the farm meeting with our apprentices, we begin by looking back at the to-do list from the previous week. We check off what we accomplished, and write the date when each task was completed. We add any relevant details or notes, such as how we amended the bed we planted, or what variety of seed we used. We then move to a new page and write our to-do list and our individual goals for the coming week.

Planting Spreadsheets

Appendix B contains more detailed explanations and examples of my spreadsheets. I use Excel for my planting spreadsheet. It includes columns for planting code (shorthand for the location and year of each crop), crop, variety, seed source, date of purchase, planting date, method of planting (direct-seeded or started in pots or trays, et cetera), quantity, bed length, germination date, germination comments, transplant date, spacing, depth of seeding, description at time of transplanting, first harvest date, last harvest date, yield, cultural practices, amendments added, and a final one for comments. I'm sure this list looks a bit overwhelming. Don't worry: I don't fill in every column for every planting. For example, I often miss recording germination dates and descriptions. For all direct-seeded crops, the columns for transplanting information are left blank, and for transplanted crops, the seeding depth is not applicable.

Generally, the more information you can get down, the better. You may look back and learn that parsley always takes a long time to germinate, and never has a very high germination rate for you, so you may decide to seed it earlier and thicker. You may see that you got a higher yield out of a bed of chard that was more widely spaced than another, more closely spaced bed. You may learn that your carefully tended, early-transplanted squash actually only produced fruit a couple of days earlier than another planting that required less attention.

Harvest Spreadsheet

On this sheet I record each sale by date, code (this corresponds to the planting spreadsheet), crop, variety, quantity, price per unit, total price, market, and invoice number.

Invoice Book

Your invoice books are the first, most basic, and legally required part of your recordkeeping system. If you're ever audited for tax purposes, this is your record of income.

Expense Spreadsheet

The expense spreadsheet is where I record all

farm expenses. I usually don't get around to doing this until field work slows down in the late fall. Then I sit down with my mountain of receipts and start entering data. You will need this information for tax purposes anyway, and arranging it on a spreadsheet that categorizes your expenses enables you to learn from it. Perhaps you will see that you're spending more on chicken feed than you are making in egg sales. Maybe you'll even get some pleasant surprises and see your irrigation expenses dropping year after year. Maybe you will learn that you're spending so much on gas for deliveries that you should consider raising your delivery charges.

And, speaking of vehicle expenses, you should keep a logbook in any vehicle that you use for farm purposes. Read the odometer at the beginning and the end of the year, and write down the mileage of every farm-related trip. This way, you can accurately calculate the percentage of vehicle use that was farm-related. You can then write off that percentage, not only of all your gas for the year, but also of all vehicle maintenance and insurance costs.

Crop-Rotation Records

This is where I record what has been in every bed in the field every year since we began farming. Each winter, when I'm planning what to plant, I also plan where it is going to go. Thus I make sure that I'm not putting plants of the same family in a place where they were recently planted, and I can keep track of what areas need to have a rest from cash crops with a rotation of chickens and/or green manures. I find it easiest to record this on bed maps.

Inputs Records

This is required for organic certification, and is a list of everything that comes onto the farm that will be added either directly to the soil or to the compost pile.

Compost Records

Compost records are also required for organic certification. These show the location of each pile and when it was started, turned, and used. I also record compost temperatures to make sure each pile has heated up sufficiently.

Certification

There is another obvious reason for keeping good records, which I have only mentioned in passing: detailed records are required for organic certification. Keeping good records throughout the year will prevent months of dread and a week of hell when it comes time to put together your application for certification (or re-certification) each

Delicata squash is popular both for its delicious sweet flavour and its convenient small size.

Chioggia beets in the sunshine on washing table, bunched and ready for the market.

year. In fact, for some farmers who use organic practices, the strict recordkeeping requirements are a deterrent to certification.

Whether or not to certify is indeed a big decision. The process is somewhat onerous, and involves some expense. For me, however, the benefits far outweigh the inconveniences. By certifying, you become part of a community that can offer you support and guidance, and the standards can be an important educational resource for good growing practices. Of course, there are also marketing benefits to certification. Many buyers want organic produce, and most grocery stores and box programs will not buy organic produce unless it is certified. You can charge more realistic (higher) prices if you are certified organic, as opposed to using other terms like "naturally grown" or "unsprayed."

Above and beyond all this, I personally believe in the philosophies of my certifying organization, and in the concept of organic certification. Sure, if all consumers knew the farmers who grew their food, certification might not be necessary, but this simply is not the case. Even if a farmer sells directly to the public, she may not have the time to discuss her growing practices in detail with every individual consumer. Also, if every customer drove out to a farm or two every week to buy their food, much of the environmental benefit of eating locally grown food would be lost due to the fossil fuel use of their cars. So, we have markets, grocery stores, delivery programs, and other ways to increase the efficiency of getting food to consumers.

Certification then takes the place of first-hand knowledge of a farmer's growing practices. Certification assures consumers that farmers, at a minimum, meet the written standards—and in fact, many farmers exceed those standards. Without certification, there can be confusion about what exactly "organic" practices are. I have heard farmers say, "Oh yeah, I'm mostly organic," and then I come to realize that they use certain agricultural chemicals that they personally do not think are harmful, but which are prohibited under certified organic standards. Last but not least, by being part of the certification process (whether as a consumer or a farmer), you are supporting a movement that researches and strives to improve sustainable agriculture.

The Business of Farming

Never forget that, whatever your reasons for wanting to start an organic farm, it is a business, and you will not be able to farm for long if you neglect the business side. You are not betraying any lofty principles by wanting to make a living at this. In the current North American economic climate, it is taken for granted that small, family farms are marginally financially sustainable at best. Nearly all farms (conventional as well as organic) are supported either by a spouse working off the farm or by government subsidies. To a large extent, our society has chosen to "outsource" the growing of its food to regions with cheaper labour and land, questionable environmental and labour standards, and further agricultural subsidies. By importing our food, we undermine our own rural economies and we participate in a system of global environmental degradation. When we support a global, commodity-food system, we sacrifice diversity, flavour, and nutrition to the supremacy of varieties that are selected for their durability in transport, and then picked unripe. Perhaps most seriously, as we lose local farmers, we compromise our food sovereignty because we lose the very basic ability to feed ourselves. This need not be the case. We can choose to take control of our own food systems, but this will require a cultural change away from thinking that price is the most important factor in making food-purchasing decisions. In this time and place, it is a revolutionary act to run a farm that is both environmentally and economically sustainable.

When starting your farm, set up good business practices at the same time as you set up your plans for growing. Hire a bookkeeper if you need one. Before you even begin farming, get a GST number (if you live in Canada). Because produce is GST-exempt, you do not have to charge it when you sell your vegetables, but once you have your number, you can get back all the GST you pay on farm purchases. This really adds up, and it only takes a few minutes to apply. Just as important as getting a GST number is getting a farm tax number and using it to avoid paying provincial sales tax (PST). In British Columbia, you get this farm tax number from the land assessment office, where they determine that the land is indeed being used for farm purposes. As well as avoiding paying PST on farm purchases, the owner of farmed land pays a reduced property tax rate.

Keep your ear to the ground for grants that might help you, and, of course, keep careful track of all your expenses, in order to write them off on your income tax. If you decide to hire people to work on your farm, you will have to do payroll, withhold taxes, and register with workers' compensation (WorkSafe BC in British Columbia). This may be the time when it pays to hire a bookkeeper. Remember that any government programs, such as tax exemptions or grants, have been put in place because governments want to encourage farming. They are not "loopholes," but legitimate ways to help farms succeed.

Woman, Farmer, Mom

I was living in Edmonton when Brian and Jane called to ask me if I wanted to farm on the piece of land that I now co-own with them. Not only did I know nothing about farming, but where I grew up, the word "farm" means something entirely different from the kind of enterprise I now run. When you say "farm" to Albertans, they picture grain, livestock, or both. Market gardening isn't really on the radar. And if you say "farmer" to an Albertan, or at least the Albertan who lived next door to my parents, he pictures a man. My parents' neighbour had heard that I was planning to move to BC.

"What are you going to do there?" he asked.

"I'm going to be a farmer," I replied.

"Oh, you're going to be a farmer's wife?"

"No," I replied. "I'm going to be a farmer."

Then, once again with nearly the same words, "You're going to marry a farmer?"

And once again, this time with voice hardening slightly, "Um, no, I'm not marrying anyone. I'm going to be a farmer."

I can't blame him. It takes a moment to integrate into your thinking an idea that has never occurred to you before, and for him, the notion of a single woman farming was just such an idea. Looking back, I am somewhat stunned that I had the—*what's the word?*—not courage or even confidence, but more like naïveté, foolishness, or even gall, to jump into something I knew less than nothing about. I think my aunt and uncle's call came at exactly the right time. I was still young enough to think I could do anything if I just set my mind to it. Just as important, I had a safety net: wonderful, loving parents who had taught me to be independent and to take chances, but who, I knew, would bail me out if I really needed it.

There was also the distant memory of the pride I had felt when, as a young girl camping with my family, my father would comment on what heavy pails of water I could carry: "Almost as much as her big brother." Ever since then I've enjoyed situations that physically challenged me, and throughout my youth I stubbornly refused to acknowledge any advantage men claimed over women, physical or otherwise.

I am glad that I can now truly say I think very little about what it's like to be a woman in a male-dominated field. I'm always a bit surprised when I'm asked about my experience "as a woman." This is probably largely because many, if not most, of the small-scale farmers in this region are women. The separate but related question, about what it's like to farm as a mother, certainly looms large in my mind, and I write more about that later in this book, but "as a woman"? It's not a burning question for me. I am a woman. I am a farmer. That's pretty much it. Or, so I thought, until I had to confront some of my own hang-ups when I hired a man for the first time.

The original apprentices and employees I worked with were women. Ilya was the first man to join the farm, but in his first season he was only apprenticing part-time. His partner, Chrystal, was my main apprentice, and Ilya was away doing other work much of the season. A couple of years later, Jeremy applied to work for me. He fit the bill perfectly; he had experience, his references told me he was a hard worker, he was committed to sustainable agriculture, and he was easy-going and fun to have around. So why was I hesitating to hire him? I had to admit to myself that I had a certain reluctance to hire a man as my full-time, main farmhand. I knew this was unethical. If the situation were reversed, and a man passed over the most qualified applicant solely because she was a woman, it would clearly be discrimination.

I hired him, and had a wonderful two years working with him. He was everything I could want in an employee, so I had to do some soul-searching about where my initial reluctance had come from. I'm still not entirely certain about how to put it into words, but I think I feared some sort of power struggle. Would he take me seriously and respect my decisions? Would he respect the scale of my farm, or would he push toward a "bigger, better, faster" mode of production that did not appeal to me? Would we have good communication and develop the warm camaraderie I had with the women I had worked with? I think some of this came from my internalized sexism, which led me to assume that a man would not take me as seriously, or respect me as much as a woman would.

I'm happy to report that I'm gradually conquering that internalized sexism. I now work with men and women, both straight and gay, as well transgendered people who define their genders outside the narrow "he" and "she" that I was used to. My life and the community of the farm are enriched by the gender diversity around here. This is especially important to me because of

Heather with sons (oldest to youngest) Jackson, Walker, and Levi.

my children. One of the great gifts of farming is that my kids are growing up around so many cool adults. They live on this shared land and some of the adults in their lives are permanent, while others come and go with the seasons. I explain to everyone I hire that my kids are free-range. I do not expect my employees to babysit, but the kids will probably come down and visit or work with them in the field. Each adult develops her or his own relationship with my children. The people on the farm are very diverse, so my kids learn that there are many different ways to be.

This really hit home one day in July. My children love a parade, and were keen to march in the Victoria Pride Parade when they heard the farm hands talking about it. A couple of days before the parade, Jackson asked me what it was all about. I explained that it was a day for queer people and their friends to celebrate who they are. He didn't understand what I was talking about. He already understood what gay and transgendered meant, but it had never occurred to him that these people were "different" in any way that mattered. I had mixed emotions explaining homophobia (because, after all, if there had never been homophobia, there would not be pride parades as we know them). I was sad to have to tell him about hatred and fear, but I was very happy to see his absolute openness and acceptance of good people without regard for the boxes adults put them in.

Farming with Kids

When we began farming, Lamont and I did not yet have any children. When our first son, Jackson, was born, things sure changed! Next came Walker, and then Levi, and we realized that farming with one son had been relatively easy. If you already have a child or children, and are thinking about farming, there are some things that you should consider.

Farming with children can be wonderful, both for the children and for the adults, but it does add certain challenges and risks. It is not possible to work in the field and give full attention to your child at the same time. A young child may be interested in weeding or picking rocks for a few minutes, but probably not for several hours at a stretch. Therefore, in order for you to get any work done, your child will have to be very independent. He or she will need to entertain himself or herself outside for

Walker and Jackson help with the corn harvest.

long periods of time. For children spending time alone (within eye- and earshot, but not right beside you) in a farm environment, there are physical risks involved. Farm kids are likely to have bumps, scrapes, and insect stings. Then there are the more serious hazards like picking up a harvest knife, ingesting toxic weeds, and even getting injured by farm equipment. You will have to assess for yourself the risks on your farm, and your comfort level with those risks. You will have to balance the small chance of serious injury against the potential everyday benefits of growing up on a farm.

You need a high degree of organization to farm with children. In the morning, you will have to be prepared with snacks, drinks, clothing, sunscreen, toys or tools, and plans for diaper changes and naps, depending on the child's age. Regardless of your best intentions, organization, and energy level, it is impossible to do as much farm work with a child as it is without. I have been incredibly fortunate to have apprentices and farm workers who love my children and who understand that part of working on this farm is working with three inquisitive, energetic boys. Best of all, I have terrific, supportive parents who come for long visits each summer and help with the farm and the kids.

Occasionally though, the stress of the work and raising the kids, added onto the hormonal rollercoaster of child-bearing, has boiled over. I still cringe when I remember one scene I subjected my friends to. It was September, and time to get the tomato plants out of the L.J. greenhouse so we could plant salad greens. Robin, Rachel, a couple of employees, and I met one afternoon to get the big job done. Jackson was two. It was hot, and I was exhausted. Not only had the farming season taken its toll, but I was nine months pregnant. I had planned to leave my toddler up at the house with Lamont, but Jackson insisted on coming down to the greenhouse. Once there, however, he wanted more attention than I had in me to give. I felt I *had* to get the work done. After all, there was Rachel, also exhausted and working away with baby Elias on her back. And I worried that Robin always had to tackle more than her share of the load since she was the only one without kids. I dug in my heels, and so did Jackson. Our fight escalated to the point where he was hitting and throwing things at me, and I was yelling at him. Then I burst into tears and Robin ended up carrying Jackson away. By this time, all the others in the greenhouse were crying sympathy tears.

What a scene for Lamont to walk into when, moments later, he showed up to see how the job was going, and ask if I needed help with Jackson. He led me up to the house and tucked me into bed. Much to my amazement, my tears just wouldn't stop, until I fell asleep. I slept for several hours that afternoon, and then at 11:00 that night, I went into labour. I think the intensity of my emotional outburst was a sign of my body's wisdom: it was changing, getting ready for the birth, and demanded that I rest up for it.

Although there are significant challenges to farming with children, there are huge benefits too. Farming is work that very young children can really be involved in. At three years old, Jackson could weed, harvest, wash and pack vegetables, pick rocks, hoe, seed, transplant, and help fix irrigation. My kids are part of a family that is working together toward a common goal. This can give them a terrific sense of accomplishment, self-confidence, and security. Children who grow up participating in a farm learn not only to grow food, but also to grasp the deeper connections of organisms in ecosystems. They understand the cycles of life and death of which all of us are a part, and they see close up the wonder and beauty of nature.

Each winter our family leaves the farm for six to eight weeks to visit relatives. One January we returned in a spell of beautiful, sunny weather, and I'll never forget the joy I felt meandering the field with Jackson. We were mostly just enjoying the smell of the air and the rich green hues of the plants. We'd stop here and there and select the food that was to make our dinner that night. Jackson snacked as we walked, and at one point he called out to me, "Mom, you should try this kale—it's delicious!" Rachel and I both farm with kids and love it (most of the time). Robin has observed our families at close range and has decided not to have children.

Help! (or Farm Labour)

At the beginning of our fourth farming season, several things happened all at once. Our new house was completed, leaving the small cottage we had been living in empty. My friends also started asking me why we didn't have apprentices. Around the same time, I started thinking that I might actually know enough about farming to provide a good learning experience for someone interested in apprenticing. Lamont was offered a great contract that we knew would take him away for most of the summer (he is a geologist), and with our first son just over a year old, we realized that there was no way I could do all the work alone.

Jackson and Walker transplanting winter squash. When they're motivated, they can get a lot done!

Farm Kids and their Vegetable-Centred World

When Jackson was barely three, he learned very quickly where to find the ripe strawberries. He made such a big dent in the strawberry harvest that Heather's farm hands were confounded about where all the ripe berries had gone. It wasn't until Heather emptied some really full diapers that she realized the volume of his grazing, and the case of the slow-ripening strawberries was solved.

When Jackson's Grade One class learned about Chinese New Year, his teacher said, "Gung Hey Fat Choy," and Jackson heard, "Gung-ho Pac Choi." He cheered in a very spirited way about one of our main crops, pac choi.

Walker was a quiet toddler with inquisitive eyes. You could tell he was thinking some big thoughts but we waited and waited for him to speak. When Heather was having some drainage done in the field we heard his very first, four-syllable word: "excavator." The children all continue to amaze us with their detailed knowledge of tractors and specialty machinery

Sometimes we think the farm kids will get a warped sense of reality from watching us spend disproportionate amounts of time doing mundane chores. During the early years, there was a lot of rock-picking being done by everyone. When Jackson got his very own toy wheelbarrow at the age of four, he pulled it up to the newly turned over soil and taught his little friend Elias, barely three, to fill up his wheelbarrow. It seemed that rock-picking was what everyone did for a good time.

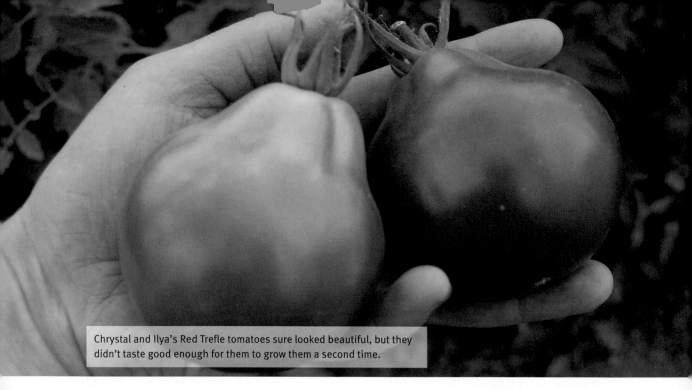

Chrystal and Ilya's Red Trefle tomatoes sure looked beautiful, but they didn't taste good enough for them to grow them a second time.

Early that spring, one of our box customers left a note saying that she was interested in learning how to farm. She lived in Victoria and wanted to apprentice part-time. I thought this might be perfect. I was nervous about teaching an apprentice, but having someone on the farm just part-time was not too overwhelming. Since she would not be living on the farm, it didn't seem like too much of a commitment, and if it didn't work out, it wouldn't be too devastating for either of us. So I met Wendy. We hit it off so well, and I enjoyed her energy, enthusiasm, and desire to learn so much, that I joined SOIL (Stewards of Irreplaceable Land), an organization that links farmers with people who want to apprentice. Through SOIL, I met Cyra, and she soon moved out to the farm. Nearly overnight, our quiet, two-person farm was becoming a little community.

I had been somewhat reluctant to bring others onto the farm for several reasons. We had

so much to learn about growing crops in those first years that I didn't think I could also learn how to teach others. I also didn't know enough to feel ready to teach. I lacked confidence and organization, and enjoyed working alone. The farm was my home, and I didn't know if I wanted strangers in my home.

Now I love having apprentices for even more reasons. When someone is working beside me, I have to question what I'm doing and why I'm doing it. I have to keep thinking and learning as I teach. I also have to be a lot more organized and efficient than I used to be. Robin once told me, "Working with someone else is great. The helper may only be able to do half of what I can do, but I can get twice the work done that I otherwise would, because I'm so much more organized!" Ideally, an apprentice is someone who is committed to sustainable agriculture, probably even planning a farming career. This commitment often gives them a level of energy that can re-inspire me, as the day-to-day work sometimes starts to wear me down. Of course, the other huge benefit to including apprentices on our farm is the labour they provide. Also very important to me is that by mentoring apprentices, I may be able to increase the number of organic farmers out there, and thus multiply the benefits of organic-food production. If we are ever going to be more than a drop in the bucket of our food-production system, we need a lot more farmers.

Since my first mentoring year with Wendy and Cyra, Northbrook Farm has continued to evolve. By 2005 (our fifth season), Cyra had moved on. Wendy's role on the farm had grown, as she worked for us half-time (for pay) and grew crops of her own in several beds in our field. That spring, Chrystal and Ilya joined us as apprentices. They moved into the cottage and settled right into life on the farm. During their first year, they worked for room and board, education, and a small stipend. They were amazing apprentices: smart, hard-working, full of questions and energy, dedicated, and serious about learning how to farm. In September of that year their apprenticeship ended and their employment began. This happened when I gave birth to our second son, and Chrystal and Ilya suddenly took on much more responsibility on the farm.

In 2006, the pattern that had begun with Wendy continued with Chrystal and Ilya: they worked for us half-time and grew their own crops in our field half-time. Since I knew I was going to be very busy with two kids and much of the Saanich Organics sales, I didn't feel able to mentor a new full-time apprentice so that year, almost all our labour was paid hourly (with the exception of a bit of help from two occasional volunteer apprentices). This took some of the pressure off me. Although we all enjoy our work more, and learn more, when we are together, I was not obliged to work alongside anyone. When I had to be in the house with the children, or on the phone with chefs, the employees would just look at the to-do list and get to work. Again in 2007 and 2008, all the farm labour was paid, with the notable exception of my dear parents who come out for a couple of months each summer. My mother takes care of meals and children, and my father helps in the field and takes on maintenance projects.

My big challenge of the last couple of years has been to become a better boss. This is a whole new skill for me. When I first started paying employees, I thought it was pretty simple: you tell them what to do and they do it. I had the idea that apprentices had a right to expect time and attention, but that paid employees would just do what they were told. With the help of a seminar on becoming a better employer, I've come to see that being a better employer gives me better employees. One of the main challenges has been to learn what motivates each employee. Does he work better independently or with others? Does she prefer to figure things out on her own or follow clear instructions? Larkin and Jeremy, my two main employees for two years, were both passionate about learning. They enjoyed taking responsibility for crops and having input in decision-making. Both could work alone, but preferred the social element of working with others. I tried to respond to those qualities in them, as well as being thoughtful and understanding about things like occasional days off. Now and then, a homemade popsicle on a hot day or coffee on a cold morning helps too. We also developed the tradition of homemade cake to celebrate any birthdays on the farm.

I am far from the perfect boss, but I think the people who work for me are generally happy with their jobs, and I have had truly amazing employees. The backbone of my communication with employees is our weekly farm meeting, when we develop the all-important to-do list. We find this meeting indispensable for exchanging information, keeping lines of communication open, and making sure we all have a similar vision of what's happening on the farm. Our agenda is much the same each week. We start by reviewing the past week's to-do list to see what we've accomplished. We then develop and prioritize the new week's list. Then, each of us shares our favourite thing from the previous week, and our goals for the coming week. The favourites and goals need not be farm related. They usually do pertain to the farm, but when they don't, we can sometimes get to know each other better, and have conversations about our bigger picture. This is important, because the kind of work we're doing, and the kind of place we're trying to create, works better when we bring our whole selves to it, not just our hands and backs.

I have now returned to including apprentices on my farm. I'm happy to get back into teaching and encouraging more potential farmers. From a financial point of view, I make as much net profit in years when I hire paid employees as I do in years when I work with apprentices. The most important thing seems to be staying on top of the timing of seeding, watering, weeding, and thinning, and this requires an enormous amount of labour.

Farming in the Future (or Tips to Avoid Burnout)

Starting up an organic farm is exciting. There is much to learn, and it is rewarding to see those first seeds sprout into life. No matter how much energy and enthusiasm you have when you begin, there will be times when you get tired, and when the length of the to-do list becomes overwhelming. I remember well standing in my field at the height of summer in 2004. Two thoughts kept running through my head. The first was, "Holy crap, I'm so tired. Why am I doing this? There are so many easier ways to make a living. I've got to quit." The second was, "Oh no, I'm screwed, because I can't think of any other job in the world that I want to do." I realized then that I was completely hooked on farming, and that I had to find ways to do it that wouldn't burn me out.

Since then, I have become a bit better at managing my stress. There are still times when it feels like way too much, and every season, Robin, Rachel, and I each have at least one tearful breakdown on a Monday night during box-packing. This happens when the work has us worn down and what would otherwise be a manageable problem (wilted pac choi, flea beetle infestation) seems like an insurmountable crisis. But those days are becoming fewer and farther between, and we are gradually learning to see them for what they are: bad days in a usually great life.

So, how do you keep the bad days to a minimum? Pay attention to your own rhythms. I'm not a morning person, so I usually ignore farmer stereotypes about being out in the field at the crack of dawn. Let's face it, at Victoria's latitude, there are too many hours of daylight in the summer

One of our home-delivery boxes packed with a wonderful variety of freshly harvested produce.

Lacinato kale (or its close relative *Nero Di Toscano*) is the least productive of our kale varieties, but it is the most popular with chefs.

for us to be working in the field from dawn to dark, so we choose to take breaks. The first couple of years, you will probably have to work fourteen hours a day, seven days a week for six months at a time. Fortunately, there's a good chance that for a couple of years you will *want* to work like this. Starting any business is incredibly challenging and exciting, and as you start farming, you will be rewarded by seeing nature respond to your actions. There's nothing like looking at trays full of glowing, germinating seedlings to make you want to skip lunch and keep planting! However, this pace is not sustainable permanently.

We all have different strategies to help us stay physically and mentally healthy. I try to take one day off each week, and now that I have children, I usually don't work outside after dinner. One farmer friend takes two hours off in the afternoon three or four days a week. During that time she rests, swims, walks, or catches up on recordkeeping. Other farmers make sure they take one "getaway" each summer, even if it is just two days. Whatever you decide, the key is to make sure it happens. You will never feel like you are able to take a day off, so you have to decide in advance and simply stop working on that day. It is especially important if you live on your farm to physically get away sometimes, even if it is just for a two-hour hike on Sunday.

One evening a couple of summers ago we met with the owner of another box program in Victoria. She had brought farmers together to talk about expanding the growing and selling capacity on the island, possibly through forming

book was almost unethical. I was so discouraged that I thought I shouldn't encourage anyone else to do what I was doing. It's just too hard, pays too little, and can be too discouraging, I thought.

The next Saturday our market table sold over $2,000 worth of vegetables for the first time, and customers thanked us for the work we were doing. The next Monday morning I was out harvesting to the sound of birdsong when the light was still golden and every leaf I touched was sparkling with dew. I was filled with a deep sense of peace and I knew that for me, there is no better job in the world than farming. Tuesday I did some research and came up with a mummy berry-containment plan. The foil pie plates I tied onto rebar stakes to deter the ravens from the melon patch even proved quite effective. Melanie reported that the diseased tomato plant tissue had been removed, and the blight did not appear to be spreading.

Now, as I write this, I am in North Carolina, enjoying a good long winter break with my husband's family and our friends. I love the winter freedom that farming gives us, but even as I enjoy the rest, I'm excited about returning to eat food that we've grown with our own hands. I eagerly look forward to planting those first seeds in the spring, because I know that each year, not only does my knowledge increase, but more importantly, my soil improves. Each year's hard work in building the soil pays off in increased soil life and health. Every year is different: different weather, different vegetable varieties, different challenges and different successes. I will never stop learning, and, as long as I am able, I will never stop nurturing myself, my family, and my land by organic farming.

a new distribution company to facilitate produce marketing. The catch seemed to be that there was just no more produce out there; most farmers were selling all they could grow. Late that night Rachel sent Robin and me an e-mail. She was fired up. "We've got to finish that book! We have to get more young people farming, and show them that it's something they can do!"

Exactly one month later I was standing in my field, having just discovered widespread mummy berry (a fungal disease) in my blueberries. On my way down the field I had seen the first raven-pecked, still-unripe melons. Then, before I could return to the house, Melanie, the manager of the Saanich Organics crops, waved me over. "Did Rachel tell you? We have early blight in the greenhouse." The idea struck me that writing this

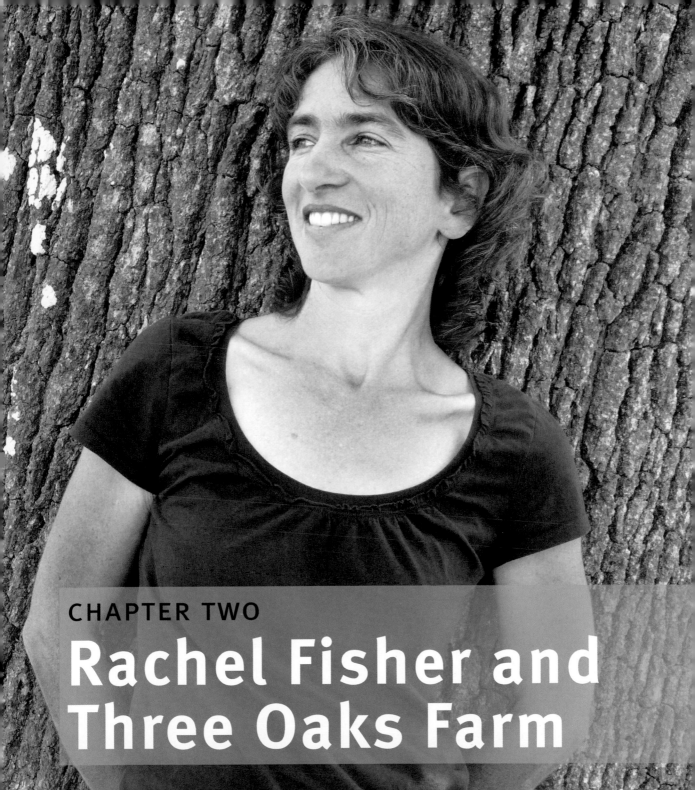

CHAPTER TWO

Rachel Fisher and
Three Oaks Farm

It can happen at any time. You may be harvesting, weeding, pulling out fence posts, digging a trench, turning compost, planting seeds, or pumping air into your wheelbarrow tire. Your cat meanders toward you and rubs up against your leg, or a flock of honking geese shoots by directly above you, or a hummingbird whirrs in your peripheral vision, and your attention is broken from your task. You look around and feel a deep appreciation for where you are. A spell is cast. Whatever doubts or fears might plague you at other times, a moment like this washes them all away, and brings a profound sense of rightness that pervades your whole being. You are where you should be. The garden around you glows with health and vitality, and the natural lines of the surrounding farm soothe your eyes and mind.

It happened to me most recently on a November morning, while I was harvesting salad greens in the middle of an enveloping fog. It was Sunday and the usual dull roar of the highway was non-existent; I was cocooned in the fog. The chirping and flittering wings of the towhees seeking seeds in the fallen, blackened sunflowers of summer caused me to look up. Other than the bird sounds, all was silent. Once again, I had that feeling of rightness, of satisfaction, and also now, a certain awareness that this feeling would come repeatedly, as long as I have the sensitivity to recognize it.

I've often been asked why I have a farm and what drew me to it. My answers are true: I am drawn to work that is ecologically healthy, that involves my body and mind, that harkens toward sustainability and thoughtfulness rather than blind, material consumption. But those answers never feel quite right, as though I'm missing something important. Now I know it's because I need to describe the sense of wholeness one feels being engaged in this work. That the small

things, like hearing the clip-clop of hoofs as my neighbour rides her horse down the road while I cut salad greens in the fog, or the scent of basil as I brush past the plants in the waning heat of a summer evening, or the act of pulling a couple of bright orange carrots out of the ground and enjoying them with my son and daughter, are at the heart of why I am a farmer.

Apprenticeship in Paradise

When this journey started for me, fifteen years ago, I was an idealist, a wannabe back-to-the-lander, with extreme environmental views and a big cynical chip on my shoulder about the excesses of Western society. I struggled with a strong need to do the kind of work that would make a positive impact on the world around me. It didn't have to be glamorous or make lots of money. I loved the idea of being in constant contact with the earth, and of living life attuned to the change of seasons.

Rachel's daughter, Jade, in 2010 with an overwintered Walla Walla onion.

For six months I apprenticed with an experienced organic farmer, Mary Alice Johnson. That summer was one of the best of my life. Producing food proved to be nourishing many times over, from seeding, to planting out, to watering, weeding, watching, and finally harvesting and eating. I was ecstatic every day, waking to the sounds of birdsong and the breeze rustling among the treetops, working a long morning, and enjoying a communal outdoor lunch with other apprentices, volunteers, and occasional visiting farmers. We worked a further couple of hours in the afternoon, and spent evenings on our own projects, cooking simple, delicious meals or swimming at a nearby lake.

Saturdays were market days. All of Friday was allotted to the harvest and we fanned out around the farm, cutting, digging, picking, washing, and beautifying produce that was already near perfect. At the height of the season the cool room would be packed full with heaping bins of peas, broccoli, potatoes, three kinds of berries, beans, salad greens, beets, onions, garlic, leeks, tomatoes, cucumbers, eggplant, radishes, summer and winter squash, basil, Swiss chard, kale, and Asian greens. We also produced herbal vinegars, bunches of dried lavender, and dried herbal tea mixes. We were up at 6:30 Saturday morning to harvest lettuce and flowers, and to pack the pickup truck for the hour-long drive to the market.

Selling the produce at the farmers' market was a good experience, but I really enjoyed the harvest days. I started my apprenticeship in April, a full month before the market was due to open. My heart felt like bursting when, after four weeks of work that consisted of some harvesting, but mostly composting, seeding, planting, watering, and weeding, we began to harvest in earnest and suddenly it all seemed to fall into place. This is what it was all for. To see containers full of beautiful, shiny, vibrant food that I had just picked made me feel fulfilled in a way I never had before. That feeling is still the same, even now. Although much has changed since those fanciful early days, I always try to honour that sense of deep satisfaction that arises at the sight of a bountiful harvest.

My apprenticeship was invaluable. Working in a hands-on environment with an experienced farmer is absolutely the best way to gain practical knowledge in a short time. My only regret about my own apprenticeship is that I did not stay for a full year. Long afterward, while running my own operation, I realized that I had no knowledge of what to do during the cold side of the calendar year. Though our winters are mild and there is much to learn about season extension and timing in terms of plant propagation, even in cold areas there are plenty of possibilities, particularly in greenhouse production.

Learning about growing crops is only one component of educating yourself about organic farming. Choice and layout of crops, care of the soil, composting techniques, pest management, weed management, machinery operation, value-added production, marketing, livestock management, irrigation, and working efficiently are of equal importance. All these things represent worlds of knowledge in and of themselves, and our challenge as farmers is to have these systems working together in harmony.

One way to get started farming with little financial input is to live simply. Very simply!

My First Market Garden

I moved on from apprenticing when I was offered the opportunity to take over a large vegetable garden that had been cultivated by friends just down the road. The situation there was a little unconventional. My friends had approached a family that owned five acres about starting a vegetable garden on their land. They were welcomed, and not long afterward were invited to build something simple to live in. Thus "the bender" was born, modelled after squatter shacks in England. It consisted of a round, raised platform, about fifteen feet across, topped by many tree boughs bent into a dome shape and tied together. A tarp was stretched over the dome and glued around a couple of windows that never proved to be watertight. Though it was a challenge, I kept reasonably warm in the winters with a tiny

woodstove that worked well if the wood was cut small enough and the fire was set just right so the pieces wouldn't collapse and smother the flames. I had an outdoor kitchen and an outhouse, and there my little cat and I lived for two years while I experimented with market gardening.

The family generously welcomed me too, and asked only for access to garden vegetables in exchange for my living on their property. They loved animals and were active in 4-H. Wherever I went outside the garden I was likely to attract a trail of four-legged friends made up of any of the three dogs, three horses, three miniature horses, goat, mule, and too many cats and rabbits to count.

At first I was stuck on the notion that I had to prove myself, that because I had always been given instruction I didn't really know how to grow crops for myself. It wasn't long before

I realized that I did indeed know; it was only my confidence that was lacking. Over time, my garden flourished and I brought all manner of produce to the weekly Moss Street Community Market in Victoria. I didn't have large volumes, due to lack of space, but I did have variety and the pride of having produced something myself, with my own labour, ingenuity, and care.

I look back with fondness on that time in my life, when I valued simplicity so highly, and enjoyed living without certain conveniences. Growing food was elemental, and fit well with chopping wood and carrying water. Much of the time I was alone with my thoughts and nature, without the responsibilities of children or the burden of major expenses. It was a blissful intro-duction to what was to come later for me, as a full-time farmer, businessperson, property owner, and parent.

Redwing Farm

The turning point for me as a farmer came when I moved to the Saanich Peninsula. A woman for whom I worked part-time offered me a lease on some land. She farmed on a separate property and was active in protecting local farmland and green space from development. She wanted to see the land being productive and wanted to qualify for the lower agricultural tax rate. At that point my commitment to organic farming bumped up a notch. I was no longer "trying it out," or "market gardening," but "farming." I made a conscious choice to be a farmer and stopped waiting for signs. Everything was unfolding easily; doors kept opening and all I had to do was step through them.

The main feeling I remember from that first year was a strong, almost overwhelming sense of determination. I needed to be successful and to prove myself in a society that scoffs at farmers' choices of livelihood, even while it complains about the price of a head of lettuce. There is a great disconnect in the minds of the general population between the price of food and the livelihood of a farmer. People seem to feel that it is their right to have cheap food; they don't make the connection that this prevents the farmer from making decent money. Farmers have to hustle, work on leased rather than purchased land, and often work at other jobs to make ends meet.

My kitchen had pallets for a floor, a wooden spool for a table, and a two-by-four to keep the horses out!

Heather with a flat of fresh picked strawberries ready for market.

Food From Afar

When people taste our strawberries and melons at the market, the explosion of flavour blows them away. Our palates have become used to bland-tasting food that has been picked unripe, transported from afar, and stored for weeks. The average store-bought strawberry is of a variety chosen for its low sugar content, since more sugar causes rot to set in more quickly. Varieties are also chosen for their ability to be shipped long distances. They have tougher flesh and skins, and do not bruise easily. Is it any wonder they don't taste great? As a chef once said to me, "It's more like the *idea* of a strawberry."

Root crops can store fertilizer and pesticide residues in their fibres. Ever notice a metallic aftertaste when you eat a raw, conventionally grown carrot? This is the kind of food we are used to, and it's due to consumers' own demand for cheap, out-of-season produce. We want zucchini, tomatoes, oranges, and melons all winter. Try eating more seasonally and locally. Your food will be bursting with life and so will you.

I could see organic farmers in the Victoria area making a decent living on small pieces of land. Compared to the average conventional farm, there was something different about the organic farm model that went beyond simply not using chemicals—something alive, progressive, stimulating, and financially heartening. I was powerfully drawn to this way of life as it called to my sense of vitality and my ecological integrity. It held the promise of being rooted on the land and producing something of value with my own hands. I dearly wanted to succeed.

Greenhouse Woes

My first major task was to build a greenhouse. Having no building experience I chose to buy a greenhouse kit rather than attempt to build one from scratch. It might appear to be easy to assemble a kit, but reading the directions was like trying to read a foreign language. Now, four greenhouses later, I laugh at how intimidated I was, but it was a very real feeling at the time. I clearly remember the day it was delivered, with all its neat little stacks and neat little boxes of nuts and bolts and other hardware. After one look at the directions, I put them right away and didn't touch them again for another six weeks. Whenever I passed by the stack of hardware that was to be my greenhouse, I felt the dread that only inexperienced builders can have when faced with what they perceive as a major project.

In the end, with the help of family and friends, I got that greenhouse up and ready in time for tomato planting in May. It wasn't hard: the directions were good, and all the parts that were supposed to be included were there. The only trouble occurred when we tried to do things the way we thought they should be done rather than the way the manufacturer suggested. Over time our mantra became, "When in doubt, read the directions." When it was finished, I realized that I had wasted too much energy being stressed over the anticipation of difficulty, and I decided to start to take things in stride and avoid overloading myself with expectations. I wish I could say I did so from then on, but I *can* say I'm getting a lot better. There's nothing like running a farm for a little personal therapy.

On a more practical note, there are two pieces of advice I will share about greenhouse-building. First, build it as long as your funds and space will allow. If you cannot afford more length at construction time, provide for a future addition in your site plan. The benefits of greenhouse growing are huge and your plans are almost guaranteed to quickly outstrip your space. Much of

the cost is in the end walls and door hardware, so adding more arches later is cheaper than building new greenhouses.

The second has to do with what type of footings you choose for your greenhouse. I'd recommend metal posts over a wood or cement base. The posts are an option you can order with the greenhouse kit and they can simply be pounded in to different depths so that the tops are level with each other. The arches fit over them and are bolted in place. This system eliminates the need to level the land, thus allowing for one of the necessities of winter growing here on the coast: drainage during the rainy season. At one time I thought that working on flat ground would be preferable, but with the volume of rain we receive here, I now believe it is best to work on a gentle slope, so the winter rains don't collect and flood your greenhouse, destroying a good chunk of your winter income.

Rachel thought of her crop rotation in terms of a wheel. Each section was like a slice of pizza; the paths ran diagonally out from the centre, and each year the family of crops in any given section was moved to the one on its right.

Bed Layout and Rotation Plan

I wanted my farm to break the norm in terms of crop layout. I scoffed at the long, straight lines that most farmers were using because they seemed to me too indicative of industrial, monoculture farms. I declared that I wanted a "garden," not a grid. To decide on a crop-rotation plan, I hit the books and found that the experts all seemed to have differing opinions regarding the best order in which to plant families of crops from one year to the next. This shouldn't have been surprising since I had often found contradictory advice when comparing information in gardening books. So I took what I felt was the best of this advice and came up with my own seven-year rotation: root crops, greens, brassicas,

nightshades, alliums, cucurbits, and strawberries. I then looked at my land to see how it might divide evenly into seven sections. Since it was square, with a ten-foot-wide grassy strip reaching from one side to the centre, it seemed natural to fan the sections out from that grassy centre, like slices of pie, or, as I fancied, like a wheel with spokes dividing the sections. The wheel metaphor had great power in my mind because every year, the crops were rotated one section, following one another in the march of the ages.

It was a lovely idea, it looked great on paper, and it achieved what I had hoped for: a beautiful garden. It even worked nicely for a while. In the beginning I curved many of my beds, starting with the short ones near the centre and working outward with longer and longer curved beds, in a ripple effect. It was a wonderfully organic design, which mimicked the non-linear character and the roundness and wholeness of the natural world. And as long as I was skipping around the garden watching dragonflies circle lazily on the sunbeams, it was great to work in. It wasn't long, though, before I got more serious about growing food and started to move from a philosophy of "I just need enough money to get by" to "I'd like this farm to really make some money." At that point all the inadequacies of my lovely round design came to light and I had to admit that those straight lines I saw on others' farms really made some sense.

Mulching and Tilling

During and after this period I was practising a no-till approach to farming, in which I used hay mulch to condition the soil instead of a rototiller. "Mulch" is a term used for any material that is laid on the soil for some or all of the following purposes: retaining moisture, suppressing weeds, adding organic matter, and protecting against erosion or frost. Mulch can be hay, straw, grass clippings, compost, bark, sawdust, wood chips, or even plastic. Straw is lovely to work with, and breaks down beautifully, but is costly on Vancouver Island. I found a good source of hay at $1.00 a bale and that turned out to be quite cost-effective.

Throughout the five years I used this method, I became convinced that mulching with organic material in a no-till or minimal-till system is the best way to nurture and improve soil life and structure, and thereby fertility. Why does it work so well? It is no great revelation, since nature has been doing it for eons. Leaves or needles fall from trees in autumn and mulch the soil all winter. They provide food and habitat, and protect the soil, tree roots, and seeds against the heavy, eroding action of rain and snow and the damage of freezing temperatures. They eventually add to the detritus left from previous years' leaf litter. Grasses also die back, after seeding themselves. The dead straw flattens, providing a protective layer over the seeds. This process repeats itself year after year, decade after decade, creating layer upon layer of beautiful, rich, living, black soil. Anyone who has seen a hay or straw bale slowly decompose into the earth knows first-hand about this seemingly miraculous process. Like most biological systems, the logic, timing, and efficiency of a natural mulch system is virtually flawless.

Garlic takes well to mulch, which keeps the weeds down, protects from frost in winter, retains moisture in spring/summer, and encourages soil life.

In the garden, mulch supports and encourages the natural processes already existing in the soil: the work done by worms, microbes, bacteria, fungi, insects, and other micro-organisms. With only a little effort from me, my soil came alive with earthworms drawn from the clay depths to the surface to break down the hay. Japanese beetles thrived, and all kinds of other critters set up shop, some recognizable to me, some not. The vast majority were beneficial. Only two seemed to cause any damage to my crops: slugs and woodbugs.

The Reign of Slugs

I had become quite familiar with slugs in my first garden in the rainforest climate of Sooke, where they reign supreme. When I started using hay mulch at Redwing Farm, they showed up in the massive numbers that I was used to. I tolerated them, even appreciated their slimy, delicately tentacled beauty as a part of my garden world. Though I did a lot of hand-picking and relocating first thing in the morning and at dusk, there always seemed to be hundreds more the next day. I also tried many of the methods that the books or other gardeners recommended, such as laying lengths of wood in the pathways. Slugs are attracted to their moist undersides and they can be turned over and picked off in the mornings. A shallow bowl of beer also attracts them and assures them a happy end. These and other methods were useful to varying degrees. Slugs love broccoli seedlings, and my favourite means to protect them became laying a thick circle of dolomite lime on the soil around each plant. Though powdery to the human eye, the ground limestone acts like glass

Ducks love slugs! What a great way to cope with these pests.

shards on the tender underbellies of the slugs, and they stayed away. This method worked very well but the lime required replenishing after each rain. Eventually all this became tiring. My garden was growing, and I began thinking about a different way to deal with the slugs. I still remember the day I found about fifty slugs dangling off the leaves of a large, overwintering kale plant like a bunch of unwelcome Christmas ornaments hanging from an unwilling tree. That was the day I'd had enough!

Duck Duties

I'd read that poultry can help to keep the slug population down, and no sooner had I set my sights on ducks than some friends offered me a pair of female Muscovies. They were the last of a larger flock that had been given away because our friends were moving. Another friend had just bought a farm and had inherited some empty rabbit hutches. I had ducks and a duck house within a week.

I built a pen with four-foot-high wire fencing and some lengths of rebar that I had lying around. At first the ducks stayed in the pen and I intended to move it around every couple of weeks, so they could clear each area of slugs and fertilize it with their manure at the same time. Well, since there is always too much to do, I didn't move it often enough, and I became uncomfortable with having a heavy concentration of manure in one place. I have also never been fond of seeing animals caged, and it wasn't long before they were ranging freely, confined to their pen and duck house only at night for protection against dogs, raccoons, and mink. I did have to clip their wings to prevent them from flying away. This is of course a form of confinement, but one with which I was relatively comfortable. Only the ends of the flight feathers are cut, which is something like cutting fingernails; it doesn't hurt if you don't take too much off.

I'd lock the ducks in the pen at certain times rather than letting them snoop around. One of those times was when I had planted out a lot of seedlings that were too delicate to handle being stepped on by duck feet, or having their leaves ripped at by duck beaks. Ducks love broccoli and other cole family seedlings as much as slugs do. Once these plants were established, with a good root system and a strong stalk, the ducks couldn't hurt them. Lettuce is also a favourite food for ducks, and my solution most years was to simply not grow it. I did have some lettuce in my salad beds but I protected those beds with crop cover.

One year I put a fence around my lettuce section to keep the ducks out. First-time visitors to the farm thought it was amusing that I let my ducks wander free but kept my lettuce in a pen.

Robin in her salad greenhouse, covering the edges of the row cover with soil to protect against flea beetles.

Floating Row Cover

Floating row cover is white, spun-poly cloth that looks like giant sheets of toilet paper. It lets light and water through, but keeps out insects, and it has become an indispensable tool on our farms. Row cover is great for excluding pests like flea beetles, carrot rust fly and cabbage moth, as well as for reducing transplant shock and preventing damage from light frosts. It makes crops grow faster in the spring because it has an insulating effect. It can even reduce wind damage and extend the season into the fall. Although we would like to reduce our dependence on plastics on our farms, we admit we are row cover junkies.

In this area we are besieged by flea beetles. These little bastards ravage the leaves of plants in the brassica family (cole crops.) Our salad mix contains many of these mustard greens: arugula, mizuna, komatsuna, kale, mibuna, choi. The leaves are most susceptible when they are young and tender. Although the damage is usually cosmetic, it leaves the salad unmarketable. After much experimentation, we've concluded that around here, the only way to avoid losing entire beds of greens is to cover them completely with row cover. We have to bury the edges of the cloth to seal the bed before the first seeds germinate. If there is even one little hole, the beetles will find their way in and multiply. There is nothing more heartbreaking than lifting the cover off a bed you thought would be perfect, and instead being greeted by a cloud of flea beetles and munched greens. At a workshop, an entomologist explained that bugs navigate by smell. To them, a tiny hole in row cover is as obvious as a glowing cinder in a dark room would be to us. When we harvest our salad beds, we uncover a bed, cut the greens, then immediately re-seal the bed. If you do get a beetle infestation in your bed, the best way to cope is overhead watering with a sprinkler over the entire length of the bed until the soil is saturated. Then, cover it immediately.

The ducks were excellent at their job. Not only did they enjoy dining on slugs (size was no obstacle), they also found slug eggs by burrowing under the hay a little. They also loved woodbugs, much to my delight. They swiftly learned to waddle-run over whenever I was turning soil with a fork, in order to not miss the smorgasbord of cutworms, wireworms, miscellaneous larvae, and of course earthworms. I didn't mind sacrificing a few earthworms and other beneficials for the advantage of being rid of slugs and woodbugs.

Since they were used to having their freedom most days, the ducks became quite indignant when they were confined to their pen. They'd march up and down along the fence wire and stick their heads through every opening, valiantly trying to escape. It wasn't long before they learned that if they climbed up one or two rungs in the fence, they'd find larger openings through which they could squeeze and drop down to freedom on the other side. (Some wire fencing has graduated holes, small at the bottom of the fence, larger at the top.) Muscovies have small claws at the end of each tendon in their feet so, in combination with some wing-flapping, they were able to get a good grip and could climb quite well for a short distance.

As their wing feathers grew back and they regained some of their flying ability, the ducks would once again start planning their escape. One would jump and flap up to the top of the duck house and survey the top of the fence line for a while, extending its neck in and out and turning its head from side to side, letting each eye get a good look at the distance to be crossed. Finally it would launch, flapping wildly, coming to a tumbling landing on the greener side of the fence. Though entertaining to watch, these antics eventually caused me a lot of frustration, and I replaced the fencing with six-foot-high wire that had only two- by four-inch openings. Not a chance they'd squeeze through those. One of them did try to flap/climb up and over it before she realized just how high it was. It was probably one of the funnier farm-related things I've ever seen: a duck climbing higher and higher, to six feet, flapping frantically, then wobbling in surprise at the top and finally tumbling down in a cascade of feathers, wings, and bruised dignity.

There have been a few instances over the years when I have used the cheapest materials, or just used whatever materials were around, as I did for the duck pen. Sometimes this works, particularly if funds are short. Quite often, though, I've later had to replace everything with the more costly materials I should have used in the first place, thus spending even more money. On the farm, as in other aspects of life, there is usually a good reason for buying the better quality items and doing things well the first time.

From Mulching to Tilling

Alas, the day came when I felt compelled to stop mulching and buy a rototiller. I had been using about a hundred and fifty bales of hay a year on my three-quarters of an acre. In the fourth year I purchased a load of hay that turned out to have a lot of weed seeds in it. They were couch grass seeds, also known as quack grass, the bane of many a farmer. Couch grass spreads not only by seed but very aggressively by root as well. The loose, moist, living layer of topsoil, created by four years of composting and mulching, presented a perfect medium for the seeds, which germinated and spread like wildfire. The roots did their thing too and within a few months it was all over the garden.

My planting method had been to remove the mulch, fork up the soil loosely, and add compost. This was always a difficult task for me because there were so many worms that I would inevitably spear some. In true no-till systems, the soil isn't forked at all, but I felt my soil still had too much clay for that. After the bed was prepared, I would direct seed, wait until the plants had germinated and put on some growth, and then replace the mulch around them. When I transplanted seedlings, I would replace the mulch right away.

Mulch is supposed to kill weeds by cutting off their light source, but that durn couch grass just kept spreading its roots under the hay, sending up shiny new green heads everywhere, no matter how much hay I piled on. When I removed the hay to plant, there would be a mass of roots and new blades of grass to dig up before I could even think of planting. Naturally, this slowed my pace right down and created a whole lot more backache. In July of my fifth year, part of my farm wasn't even planted and it was because the thought of dealing with the couch grass was too discouraging. I then purchased a rototiller, and with great relief tilled the offenders in and watched them disappear.

Thus began my love-hate relationship with the rototiller. It did a lot of back-breaking work quickly and could be used to incorporate compost or other organic matter. With the hiller-furrower

attachment I could quickly make a trench for planting or drainage, or create raised beds. In short, a lot of area could be prepared for planting much faster than with my old method. On the surface, the cultivated soil presented a tidy, uniform, perfectly powdery planting medium. Too perfect. This effect came at great cost to the soil. In short order, worm activity decreased dramatically and the grasshopper, beetle, and insect larvae that I had been regularly discovering tucked into tidy little holes in the soil were all but gone. There are millions of micro-organisms in a handful of soil. I am not a soil biologist but I could see clearly that the soil structure was gone and most likely the organisms were gone too.

Due to the action of the tines, tilling also creates a hardpan underneath the tilled soil. This decreases the natural ability of the soil to let excessive moisture percolate down and creates an obstacle for organisms attempting to navigate the area; crops with deep root systems may have trouble penetrating through the hardpan to reach the nutrients below. Tilling creates a system that depends on itself. The tilled soil is subject to compaction, and overhead watering exacerbates this problem. With the natural structure of the soil destroyed and few organisms present to create air tunnels and pockets, it becomes necessary to till again to aerate the soil so that seeds can germinate and easily set roots.

Since I abandoned mulching as a management method and began tilling the soil, I have been sensitive to how much I use the tiller, doing my best to keep it to a minimum. I prefer to allow the soil a little chunkiness rather than striving for a perfectly smooth surface.

Over the years, I have been searching for the

ultimate machine, one that can do the work of a rototiller but won't cause as much damage. There is such a beast as a walk-behind spader, which would be perfect for a farm my size. However, my research revealed two problems: one, they are quite heavy in the rear, which makes for difficulty in manoeuvring. A counterweight could be placed upon the front, but this would make it even heavier which is not the most appealing idea. I also don't like the idea of having to fix something that is brand new. The other issue is that they bounce a lot, due to the up-and-down motion of the paddles, and this can be jarring for the operator. Though I haven't ruled it out as a possibility, I will keep watching for a better design to come on the market before I consider buying one.

Coping with Exhaustion

Around year four I nearly quit farming because I felt I was putting out too much energy for too little return. I was discouraged, mentally sluggish, and tired all the time. I didn't realize that it was not only the hard work that was sapping me, it was the state of my health. I had always taken it for granted that my health was good. After all, I was young, physically active, breathed clean air, and drank clean water. I ate well and I had never had a history of health problems.

When I became pregnant with my first child, my blood was tested and found to be quite low in iron and extremely low in Vitamin B12. I had been vegetarian for about ten years but that was not in itself the problem. When I cut out eggs and dairy as well, without compensating for the nutrients I was missing, my body began to use

Kat putting the finishing touches on the potato hilling. We first tilled the pathway with a hiller-furrower attachment on the rototiller, which created a furrow between the rows and pushed the soil onto the beds.

its reserves of B12 and I was heading for a major health crash. My midwife recommended B12 shots to change the situation as quickly as possible. I started to call them "smart shots" because my head would clear and I could articulate my thoughts properly again.

Rachel did a lot of work with a child on her back. She avoided using machinery during those times, but she occasionally just had to get some rototilling done.

Exhausted, and a Baby to Boot

The problem still wasn't entirely corrected by the time Elias was born, and the ensuing weeks and months of breastfeeding and sleep deprivation didn't help me catch up on my health. At this time Robin, Heather, and I were building and working in a greenhouse that we had purchased together and planted for Saanich Organics. Every Sunday afternoon we got together to plug away at the work. I recall absolutely dragging myself to those sessions, with no energy and definitely no enthusiasm. I worked with Elias on my back and though I loved the bonding and

the "rightness" of it, the physical exertion took its toll. Like many women after they have had their first baby, I thought I should be capable of doing mostly what I had done before. In reality, you are mostly doing something different—caring for a child—and you can do only a little of what you were capable of previously. As the baby grows, you get more and more of that capability back and in retrospect, it really isn't too long before you are back to doing mostly what you did before.

Our son finally started sleeping properly when he was two years old, and, happily, the effects of sleep deprivation faded away. Not long afterward, I stopped nursing and I was able to get my B12 and iron levels back into balance. I had energy again, and as Elias became a little more independent, I could throw myself into the work once again. Robin and Heather had been incredibly supportive during the whole ordeal, particularly Robin, who would come over to look after the baby so I could grab a nap here and there. Sometimes, and I'm not kidding, those naps were the difference between sanity and insanity.

Looking for Chocolate Cake

After seven years at Redwing Farm, my partner, Grant, and I started looking to buy land. We were a family now and we wanted more stability We were fortunate to have family finances to draw from which would allow us to buy in our over-priced area. There were lots of details to consider in our search but we had five main criteria: price, location, land, water access, and housing.

My ideal piece of land would be fertile, with

a good water supply. Year-round sun exposure was important, and I was looking for a gentle, south-west- or southeast-facing slope, in that order. I hoped it would not have chemical contamination, and I preferred a tranquil setting with natural surroundings. I would have used just about any type of soil, as organic farmers tend to do, but naturally I hoped for soil that had inherent structure and good drainage, and was relatively free of rocks. Having worked with rocky soil when I was an apprentice, I knew that for every hundred rocks picked, it seemed as though another hundred lurked below the surface, waiting for the next tilling to bring them up.

For over a year we scoured the real estate listings for an appropriate property in the area and looked at every one that could possibly be coaxed into something resembling a farm. We always brought along our trusty shovel, and our realtor was tickled to see us march past the house without much more than a glance, making a bee-line for the empty field beyond. After having dug holes all over the Saanich Peninsula, I was starting to think the soil was the same everywhere: rocky clay. I was amazed when we found a place that had twelve inches of rock-free, black, loamy-looking topsoil. It was fall and the soil was wet so it looked like chocolate cake and seemed to

be the stuff of dreams. Suddenly I couldn't stop grinning and I knew at that moment that I really was a farmer at heart. This piece of land soon became Three Oaks Farm.

Water

When we were searching for land suitable for farming, water was one of our primary considerations. We don't have municipal water in our immediate area, so we were happy to note that our well was a hundred and fifty feet deep and had a flow rate of twenty-five gallons a minute, which was more than adequate for the needs of the house and the farm. First I had to determine how to get water from the well to the cultivated part of the property, some five hundred feet away. Once the water was accessible in the garden area, I had to decide what kind of system I would use to irrigate the crops.

I was advised to use one-and-a-quarter- to one-and-a-half-inch pipe to cover the distance from the well to the garden. This is a rather large-diameter pipe but necessary for such a long distance, due to the pressure loss from friction. A smaller pipe would also have decreased the volume of water, thus limiting my options. I wanted the flexibility of overhead or drip

Three Oaks Farm is a short jog from the house and is nestled in a peaceful spot surrounded by forest.

irrigation, and I wanted to be sure that lack of water would not preclude the possibility of future expansion. I had the advantage of running the line downhill, which would help compensate for friction loss. In the end, I settled on one-and-a-half-inch black poly line that I buried at a depth of two feet. Though I asked lots of questions, I really didn't have much information on the well's capacity. I was really crossing my fingers that I was making the right choices and would not be compromising the water needs for the house. In the end, everything functioned just fine and my choices did prove to be sound.

Until this system was in place, I used a series of garden hoses that ran from the house to a sprinkler, duct-taped to a pitchfork, which could easily be moved around the garden. Due to the long distance and the tiny hose diameter, this proved to be woefully inadequate for the area I was attempting to keep watered, and, until I could get a tradesperson out to connect my waterline to the well I jumped around like a little frog, moving that sprinkler every forty-five minutes or so to the most needy areas.

Since it is crucial to keep seedbeds moist to achieve good germination, I lost several of them for lack of water, and had poor germination in others. Spring crept into summer and I continued to struggle, cursing myself for not having lined up a tradesperson much sooner. We happened to be in the midst of a trades crunch at that time, when all kinds of construction and other projects around the city were stalled due to a lack of skilled workers. One day my neighbour was out cutting hay and noticed my little sprinkler working away, watering its fraction of the garden. He shook his head and came over, dragging a one-and-a-

quarter-inch waterline fitted with four garden hose connections, and two tall sprinklers welded to metal wheel hubs at their base that could be moved from place to place. He then generously hooked this up to his own well. That evening we turned the water on, and with reverence I watched those sprinklers delivering their sweet goods to the thirsty plant roots. The water droplets sparkled in the golden rays of the setting sun and the *chuck, chuck, chuck* sound of the sprinklers was a symphony to my ears.

Overhead Water or Drip Irrigation?

Not long afterward, someone finally came to connect my waterline to the well and I was off and running. I had set up a drip-irrigation system for my strawberry section and for the tomato plants in my greenhouse but otherwise used an overhead sprinkler system. I find myself rationalizing my use of overhead water, since it is more wasteful than a drip system: it is simply the easiest and most economical way to water a sizeable area planted primarily in annual crops. It can be put together quickly, with little of the frustration that comes with drip systems due to their many parts, leaky connections, and the inevitability that you must visit the irrigation store yet *again* for a part you didn't realize you needed. I spent $400 for the drip system on my strawberry section, which covers an area eighty by seventy feet. This is about an eighth of the total area that I wanted to irrigate. We can extrapolate that it would cost roughly $3,200 to set up a drip system for one acre, as opposed to roughly $400 for an overhead-sprinkler system for the same area.

There are two other distinct advantages the sprinkler system has over the drip system. Drip

Robin's drip lines work wonderfully on rows of french crisp lettuce.

line is best moved off the field and stored for the winter. With anywhere from one to three lines per bed, plus the mainline, this makes for a lot of work, both removing them in the fall and returning them to the field in the spring and summer. The second point is that an overhead system uses far less plastic than a drip system. Though organic farmers do rely on plastic, which we use to cover everything from greenhouses and compost piles to crops, many of us also despise the stuff. From the manufacturing process to the landfill, it causes pollution and increases our reliance on the petro-chemical industry.

All this being said, there are some excellent applications for drip irrigation. If water is in limited supply, it is the only way to irrigate. If you live in an area where water is costly, then the price tag on the irrigation system may be lower than the final expense of your water consumption. It is far more useful for perennial crops which tend to grow large and bushy. Drip does not compact the soil the way overhead water does; the farther the water is thrown and the harder it hits the soil, the more compacted the soil becomes.

Drip irrigation is also quite useful in a greenhouse setting, for two reasons: first, with a sprinkler system, a lot of water hits the plastic sides and waters the edges, where weeds get established and are difficult to remove; second, greenhouses by nature are more humid than outside and can act as breeding grounds for fungal diseases. One of the notorious fungal diseases in our area is late tomato blight, which can ruin a whole crop of tomatoes in a very short time, if the right conditions exist. Keeping the moisture at the roots of the plants and off the foliage is crucial.

Growing Tomatoes—Rachel

I've had a love affair with tomatoes since long before I started farming. They were the first thing I tried to grow, in pots, before I even had a garden. I wasn't terribly successful at first, though the plants did their best. I was missing a few key pieces of information.

I now start the seeds in the house in late February or early March. They can also be started under lights and/or on heating cables. In our area they can be started in an unheated greenhouse in March but it can take some time for them to germinate, and a cool spring can cause a lot of worry. I scatter a few hundred seeds in a tray, but if you have the space, starting them individually in soil blocks or plug trays is beneficial. Plastic covers can help create humidity to speed germination.

Once most of the seeds have germinated, I move the trays to tables in the greenhouse. If they are kept in the house too long in the constant temperature, they grow spindly and the stems are weak. When they have two sets of true leaves, I pot up each seedling into four-inch pots. It is important to set them deep, so that the surface of the soil is just below the first true leaves. It's okay to bury the original pair of "false" cotyledon leaves since they are going to die anyway; tomato plants will root anywhere the stem touches soil, and the plant will be stronger and less leggy if encouraged to root at this point on the stem. Though consistent watering is important at this stage, overwatering can cause the fragile stem to rot. Bottom water is beneficial for a few weeks while the plants strengthen and set up good root systems in their pots.

If their permanent place will be outside in the garden, the plants should be brought outside to a semi-shady or shady place to harden off a few days before their planting date. We plant out in May on the coast. They like full sun. It is possible to grow them in partial shade but they will take longer to produce fruit. I plant mine in the greenhouse for the extra heat and humidity. The greenhouse also provides important protection from fall rains, which can cause a fungal disease called late tomato blight. Plant them at least eighteen inches apart, two feet if you have the space. I like to use three-foot beds, and I plant two rows per bed, staggering them in a zigzag pattern. I dig a hole for each, throw in a shovelful of compost, and mix the compost with my soil before planting. As before, plant deeply, to the first sets of leaves, and water in.

Now comes the real work! Indeterminate types of tomato plants will vine all over the place if left unchecked and will need pruning and staking. If you dislike this task, grow determinate varieties. They grow in a bush shape, and can be left unpruned, though they generally need staking. Determinates will produce one main crop, and a few more over the season, whereas indeterminates will produce on and on as long as conditions are good.

Pruning involves cutting off the "suckers" that grow where the leaf branches meet the main stems. There are different schools of thought on pruning: Some say to prune down to one leader to encourage more fruit production while others advocate for less pruning, claiming that the fruit tastes better if there is more foliage on the plant. I take the middle road, pruning down to three leaders. This gets the plant under control and allows for some air circulation. In hot summers, the fruit can actually get sunburned, so the extra foliage offers some protection.

Instead of staking in the traditional way, I learned from other growers to use strings that I tie on to the purlins that run the length of the greenhouse. I wind the other end of each string around the base of a plant, and wind the leaders around the strings as they grow. There is no need to tie the strings around the plant; winding is enough. If your plants are outside, you may set up a trellis system with rebar and wire, or use wooden stakes. Your string or stakes must be strong, or the weight of the plants when full of fruit will break them or pull them over. In the greenhouse, I use baling twine, and untie the lengths every year when the plants are pulled out. They are then grouped together and tied high up out of the way till the following year, when they are reused.

Tomato plants like plenty of water until a lot of green fruit has set and the first ones are ripening. At this point too much water will cause watery fruit, so I cut back, though it's important to continue to water consistently. A feast-or-famine situation will cause the fruit to split.

At last, if you've started early, you'll be rewarded in July with some delightful juicy tomatoes. The cherry tomatoes produce first, and for a week or two nothing makes it to the sales list as we get our fill of snacking while the fruit is still warm. For the next few weeks we harvest and continue pruning. Though tomatoes are a lot of work, they definitely pay for themselves. They are usually in my top three for income when I add up my crop sales at the end of the year. Around late September, at the end of the season, all the green fruit of a respectable size is picked and brought indoors to ripen in a dark, dry warm place. We enjoy the fresh fruit till December and freeze some in plastic bags to cook with for the rest of the year.

Other fungal diseases that are exacerbated by excessive moisture and lack of air circulation are botrytis (grey mould that causes damping-off in seedlings), downy mildew, clubroot, rust, scab, verticillium wilt, white rot, and a host of others with which we are pretty familiar here on the damp coast. If it's necessary to use overhead water in the greenhouse during the moist times of the year, be careful to do it early in the day to give the plants plenty of time to dry off before the cool nighttime conditions set in.

The Lay of the Land

My other hopes for potential farmland had to do with the lay of the land: sun exposure, the direction the land faced, and its relative flatness. Some shade on the growing area can be an asset even if the farm is intended mostly for summer growing. Several crops, such as spinach, radishes, mustards, lettuce, Asian greens, and peas, prefer cooler conditions and thrive in partial shade. As far as spring, fall, and winter growing go, though, it can put a real damper on your operation. Winter crops will grow slowly if at all, snow and rain take longer to melt, evaporate, and drain, and the soil takes longer to warm up in the spring. This sets back the first cultivating date. My field is in full sun in the summer when the sun is high in the sky, but due to forest along the south and west boundaries, there is a fair amount of shade in the winter. In the years when we get snow, I watch it disappear much faster at Heather's Northbrook Farm where the south-facing slope is exposed to full sun.

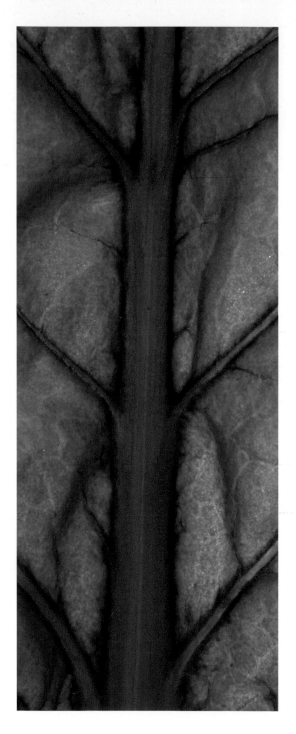

One of those moments when we wonder . . . "why do we do this?"

Winter Farming

Though we are blessed with a climate mild enough to grow and harvest food all winter, the reality of working out in the weather is far from rosy. Days are short and can be cold, windy, and rainy. On those days nothing short of donning a couple layers of clothes and stepping into full rubber rain gear will do. Hands are encased in warm gloves inside rubber gloves and we can still get near frozen fingers. If our water lines are frozen we generally have a short window in the middle of the day when they thaw out long enough to wash our muddy produce. Sometimes everything has to be moved to the hose at the back of the house and washed in a portable sink. We are unable to drive directly to our washing area, which means wheelbarrowing everything through the mud. Snow adds an extra layer of interest, and where did all those enthusiastic volunteers of summertime go?

On the bright side, when we all come together in Heather's garage to pack up the orders, the sense of commiseration is strong. Having a dry, heated space to work in for the rest of the twelve-hour day is enough to move you to tears. The produce looks extra shiny and vibrant, and we know we are coming through for our customers. The sheer variety of food we can offer is motivating: beets, turnips, rutabaga, carrots, winter radishes, leeks, kohlrabi, broccoli, cabbage, collards, kale, chard, salad greens, spinach, cilantro, arugula, pac choi, lettuce, stored squash, potatoes, onions, and frozen berries and tomatoes from summer. Greenhouses allow us to extend the season and get a jump on spring production. The large demand for local produce in the winter, and of course the additional income, carries us through the hard times. And we know that whoever is still working beside us in the winter months, frozen fingers and all, is truly passionate about growing food.

On the other hand, though the forest casts shade in the winter, it also provides wildlife habitat and brings ecological diversity to the edge of what is essentially a flat, open field. Though my farm is organic, and I seek to plot a course that appreciates and encourages the natural world, it is only within the confines of "agriculture," which is, in essence, completely disruptive of that natural world. The surrounding forest is a remnant of what existed before the area was cleared, and thus provides an aesthetic and psychic reminder of the previous incarnation of the land. Much as I depend upon cleared, open land for my living, I love the forest's presence, and it fulfills my hope for a beautiful, tranquil, natural boundary and buffer between neighbouring properties.

The Four Directions

In the northern hemisphere, south-facing slopes provide the best growing conditions in terms of heat. If the land is flat or north-facing, the sun's rays are diffused, because the area they cover is larger than if the rays contact the land at a more acute angle. The intensity of the sun's energy is increased with the more acute angle. In the summer, this is fabulous for those crops that love heat. In the fall and winter, the extra heat units can help crops continue to grow and stave off frost a little longer. In the spring, the land dries out and warms faster, and enables significantly earlier cultivation than on north-facing slopes.

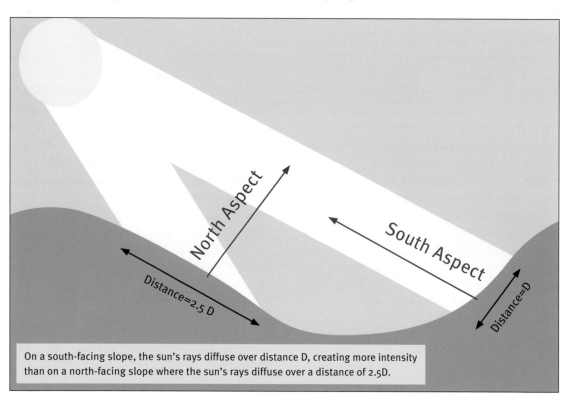

North Aspect

South Aspect

Distance=2.5 D

Distance=D

On a south-facing slope, the sun's rays diffuse over distance D, creating more intensity than on a north-facing slope where the sun's rays diffuse over a distance of 2.5D.

I watched this situation for years at Redwing Farm, which was north facing. Heather and Lamont, on their south-facing slope, were consistently tilling and planting three weeks or so before my ground was dry enough to till. This isn't to say that north-facing land is undesirable or too difficult to farm, or that heat-loving crops won't do well there. It is simply to say that south-facing land can give you a definite edge during the main growing season and your crops can continue to produce for you well into the normal "down" season.

Deer Fencing

One of our first tasks was to put up deer fencing. Having heard plenty of stories about deer busting through inadequate fences to feast all night on strawberries, peas, lettuce, and other delectable goodies, I wanted to do it right the first time. When Robin started her farm, she had an existing four-foot-high fence. She bolted two-by-fours to the existing posts and attached four-foot-wide plastic netting, which brought the total height of the fence to eight feet. The plastic netting is cheap and nearly invisible, but it gets knocked down by snow and falling branches. By the time she fenced, the deer had already discovered her produce, so she has continued to battle persistent deer.

I wanted an eight-foot-high fence, but when I sought the advice of our friend and long-time fencer, Greg, he suggested a high-tensile-wire game fence that was six and a half feet high. I was apprehensive, but he assured me that it would work, and he was right. I suspect that if I had had a garden here previously that the deer had already discovered, they might have jumped over my

This is a smaller version of the post pounder Rachel used for her deer fence. The pounder slides over the post, in this case rebar, and is lifted and dropped down to pound the post into the ground.

fence to the rewards they knew lay within. This has been the case at Northbrook Farm, where Heather and Lamont built their fence to the same height as mine, but only after years of deer trouble. The deer knew what they were after, and were willing to breach the fence to get it, whereas at my place they have had no motivation to try, and the presence and scent of our dog helps a great deal.

For the posts I chose ten-foot-long steel T-posts. Since two feet would be pounded into the ground, the posts would reach to a height of eight feet. This of course was one and a half feet more than necessary, but I wanted to allow for more height against the possible need to raise the fence in the future. If the deer did indeed jump six and a half feet, I could string wire along the top of the fence to eight feet and be assured of keeping them out.

Making the most of greenhouse space. These salad greens were planted September 1, and are ready for cutting in the third week of the month. At the end of the month the tomatoes growing in the greenhouse will be removed and the house pulled into position over top of the greens, where it will stay for the winter.

When it was time for building, Greg loaned me his post-pounder, an essential tool for the job. It is basically a heavy, hollow steel tube blocked at one end. The tube is slid over the post till the top of the post hits the blocked end. Then the pounder is raised by hand and dropped as many times as it takes to pound the post to the desired depth. Since I did most of the job on my own, I put masking tape at the two-foot mark (the depth to which I wanted the post in the ground) so I could see it from the top of the ladder.

Midway through pounding I checked each post for straightness with a level, manually pushing it straight if necessary. This was repeated at the end of the process. Since it was spring, it was perhaps the best time of year for this job; the

soil was wet, but not saturated, so the posts were easily set. I worked in hundred-foot lengths, running my measuring tape from the first post to the last, and stretching it quite tight. I then filled in the posts between at each ten-foot mark.

The measuring tape must be tight to ensure a straight line. The fence is attached to the posts in three places with pieces of wire that act like twist ties. If these ties are not tight, grass and encroaching blackberries or brush will grow through the fence and gradually pull it down.

Be sure to plan ahead as you are unrolling your fence wire. Our rolls were extremely heavy and they took two fit people to manoeuvre. As we unrolled the wire along the line of posts, we encountered a gentle uphill rise and realized, after collapsing in exhaustion, that we had started in the wrong direction!

For a final note on fencing I would advise enclosing as large an area as possible, even if you are not planning to use all the land in the near future. When you are ready, you'll be glad the work is done! If you take the other route and fence a smaller space, in order to expand you will eventually have to both add more fencing and remove some of the original fencing to create one large area. It's a lot of extra work that no farmer needs. Other scenarios may pop up as well, such as another farmer looking for growing space or pasture for their livestock. In addition, doing a lot of fencing at once means you are more able to negotiate a cheaper price with the materials supplier. Our fence cost a total of $3,500, roughly half for the wire and half for the T-posts, and it encloses two and a third acres.

Greenhouse Dream

It has been my dream for some time to have movable greenhouses. Greenhouses are a large capital investment and many of us make do with just one or two, so it seems as though greenhouse space is always at a premium, since we use them both for starting seedlings and for growing crops in the ground. The thought of being able to maximize this space without actually building another greenhouse is very attractive. There are several ways one can go with this. For my current needs, I move the greenhouse twice a year—at the end of September and again in May. This timing corresponds with my need to hire a tractor operator to do some tilling on sections of the farm, so I have him or her pull the greenhouse to its next

position at the same time. In the fall, the greenhouse is moved over an area where I have already established a crop of salad greens. Since I grow tomatoes in my greenhouse in the summer, this allows me to leave the tomatoes in the ground up to four weeks longer than I used to when I had to plant my winter salad greens in the same space. With crops that have a short maturation time like salad greens, there need hardly be a ripple in the steady flow of greenhouse income.

In February it is warm enough (and the soil in the greenhouse is dry enough), to pull out some of the salad greens and plant early crops. Good candidates are carrots, beets, radishes, turnips, spinach, Asian greens, lettuce, green onions, and peas. In May, when it is time to plant out the tomatoes, the greenhouse can be moved to the new tomato area. By then, the early-seeded crops will prefer the cooler temperatures and are established enough to handle being outside. The great bonus in this system is producing the first spring beets and carrots on the local market. This counts for a lot at our farmers' market, where the customers come early to see what the different farmers are offering, and line up where the pickings are best. The spinoff for that farmer is that the customer who originally lined up for, say, the first peas, also buys other veggies. The booths at our market that consistently offer early crops or crops out of season are also the ones that consistently sell out first, and those farmers' technique for producing these items is to maximize greenhouse space.

Another motivation for me to develop a movable greenhouse system was the need to rotate my crops. Crop rotation is one of the most important tenets in organic systems, since it helps discourage disease buildup and pest problems, and alleviate mineral or nutrient deficiencies. When I had only one stationary greenhouse, my tomatoes were in the same place year after year. I was never entirely comfortable with the lack of a rotation plan, and nor were the organic certification inspectors.

Now I have three locations for my movable greenhouse so my tomatoes will have a three-year rotation. Obviously, the more locations one can create, the better the rotation can be. The additional advantage to this system is that when not covered by plastic, the land will be exposed to the cleansing influences of wind, rain, sunshine, and freezing temperatures.

An excellent reference for information on movable greenhouses is Eliot Coleman, who writes about them at length in *The New Organic Grower* and to a lesser degree in *Four Season Harvest*. He includes both building plans and crop plans. There are many possible designs for movable greenhouses depending on the needs of the farmer and the equipment at hand. Some are built of wood, some of metal. Some roll on permanently installed tracks that run the length of all the stations, while others may be moved either by hand or with the use of a truck, tractor, backhoe, or winch.

Designing my Movable Greenhouse

Engineering is really over my head so I consulted with a friend on how to proceed. Mike is an electrician by trade but also a skilled carpenter, general handyman, and problem-solver. This was just the kind of thing that got his creative juices flowing, and flow they did. Before I knew it, two forty-foot I-beams had arrived in our driveway and a student welder was hired to come and cut

all four lower corners and weld them up into a ski-tip position. Thus the structure could be pulled in either direction, and the ends would ride comfortably along the surface of the ground rather than digging in. Since I had opted to use my existing twenty- by forty-foot metal-arch greenhouse, we simply needed to change the wood base to something more appropriate for transport. Angle iron was a good option, and less expensive, but we chose I-beams because the metal base plates for the arches could then be well above the ground and away from contact with dirt and moisture.

After the I-beams were painted to protect them from rust, we pre-drilled holes for each base plate. We then proceeded to rebuild the greenhouse on top of its shiny new skis. The next challenge was to rebuild the end walls so that they could be flexible enough to move along with the structure. In the end, we decided to simply cover the ends with plastic, securing it to the arch and to a horizontal two-by-four partway down the arch. When the greenhouse was in position, we allowed the plastic to hang and fold over itself on the ground, securing it with cement blocks that could easily be removed at the time of transport; the plastic would then drag over the crops during the move without doing much damage.

We later created a roll-up system for the front end by rolling the plastic onto a twenty-foot length of metal pipe and securing it to the pipe with plastic clips that are readily available in most seed catalogues. A crank handle was attached to one end of the pipe, and we can now roll the front end up to a height of eight feet, bringing welcome ventilation to sweaty August tomato harvesters!

How to Put Plastic on a Greenhouse

When you order a greenhouse, the roll of plastic that comes with it is designed to unroll perfectly on the ground along the length of the greenhouse. Once you cut it off the roll, it unfolds like an accordion as it is pulled over the top of the structure. Be sure to leave a minimum of a foot of extra plastic at each end of the greenhouse so you have something to work with when attaching it to the end arches. Excess plastic can be cut off later.

Tie lengths of rope to the plastic every twenty feet or so using rocks, Scout-style. (Wrap the plastic around the rock and tie the end of the rope over top of the plastic, around the rock.) Throw the other end of the rope over the top of the greenhouse so it lands on the ground on the opposite side. For a hundred-foot greenhouse you'll need six ropes, and ideally six people, to pull it over the top and hold it in place until it is attached. Another person with a stepladder and a long-handled tool can help troubleshoot inside the greenhouse if anything gets caught up on the arches. Once the plastic is on the other side, check that it is square and start attaching it to the arches. Most pre-manufactured greenhouses come with metal spring-lock that holds the plastic in place in the channels that you have already attached to the greenhouse structure. If you have built your own greenhouse, these parts can be ordered separately. Warning: do not try this on a windy day!!

Friends are people who turn up to help get the plastic on your greenhouse!

Rachel's twenty-by-forty movable greenhouse, ready for moving in May, wearing its transport hardware. The spinach and Japanese turnips were planted February 1 in this protected environment and are now able to grow without protection. April and May are slim-picking months, so having these crops ready to harvest is a real bonus.

Front corner detail showing one of the four posts welded to each corner of the movable greenhouse, the extra board attached to the front beam to add strength, the cable with turnbuckle to tighten it, and the cable on the left, which is one of two to which the tractor is hooked for pulling the greenhouse to a new location.

Supporting the Greenhouse during Transport

The main question that came up when we were discussing the movable greenhouse idea was this: how would we keep the relatively flexible greenhouse structure square? Mike came up with a simple two-tiered plan: to support the structure across its width on either end while providing diagonal support from the front to the rear corners. We had eighteen-inch metal posts welded to each corner of the greenhouse in a vertical position. We purchased two twenty-foot-long, two- by six-inch wooden beams, and bolted metal brackets onto both ends. To these brackets we welded hollow metal pipes, which slide down over the metal posts. When we are preparing to move the greenhouse, these two lengths of lumber are placed in position; they strengthen the front and back, and help to keep the greenhouse square. The one on the end of the greenhouse that is being pulled has an extra length of lumber bolted to the top, which helps mitigate the stress of the inward tension caused by the pulling.

We used cables to support the greenhouse diagonally. We made loops using cable clamps on each end of the cables, which are easily slipped over the vertical posts at the corners. They run from the front corners to the opposite rear corners. Turnbuckles help tighten them once they are in place. It is important that these cables are equally tight or the greenhouse may pull too much to one side, as we found in the first run. When set up properly, though, the cables and the lumber do a wonderful job keeping the greenhouse square, and they fit the bill for simplicity.

To provide anchorage for the greenhouse, we used a length of chain attached to a piece of aluminum buried at a depth of three feet. These anchors were placed at each corner of each station to which the greenhouse would be moved. A foot of chain emerges from the ground, which we attach to corresponding lengths of chain welded to the bottom corners of the greenhouse.

The Cost

I spent about $1,000 for materials and labour. This doesn't include the greenhouse itself, which originally cost $3,000. The I-beams accounted for roughly half of the $1,000, so seeking out a cheaper option like angle iron may be worth it. To put that cost into perspective, in the first few months of its use, from May to the end of September, the tomatoes grown in the greenhouse grossed $4,400. In addition, as mentioned earlier, I was able to start my salad greens in the next greenhouse position and move the greenhouse over top of them just as they were ready to cut.

Starting Seeds

Greenhouse space is indispensable for starting seeds before the ground is dry enough or warm enough to be worked. For years, I have used only a non-heated greenhouse, though other farmers have heat benches and/or light benches as well. In the last couple of years I have started to use my home as a warm place to start tomatoes, eggplant, and peppers, just to ensure they germinate nice and early. In our area, it can take four and a half months of care before there is a single ripe tomato, eggplant, or pepper to reward you for your labour, so the earlier the plants start producing, the better. As the weather warms, I gradually move these trays of seedlings to tables in the greenhouse, where they can get acclimatized to outside air and enjoy a greater amount of light.

One of Rachel's favourite eggplant varieties, Zebra.

To Eat the Extraordinary Eggplant—Rachel

Every gardener has a favourite plant that gets fussed over and given special treatment, even when there are much more important things to be doing! My personal favourite is eggplant. It is a bit of a challenge to grow on the coast because it prefers more reliable heat and humidity. It's important to start the seeds early, in late February or early March. I use a soil blocker to make two-inch blocks, putting one seed in each. (Four Block Maker, available from Lee Valley Tools or Johnny's Selected Seeds.) I start them in the house, though in a greenhouse under lights or on heating cables is also adequate. When the true leaves form, I transplant to four-inch pots and move them to the greenhouse. In a cool spring extra protection is necessary, either with supplemental heating at night or row cover overtop to protect from frost. If planting outside, remove the plants from the greenhouse and harden them off after all danger of frost has passed. I plant them out in late May or early June. They benefit greatly from row cover or a cloche to warm the soil until they are established and summer heat is reliable. Space them eighteen inches to two feet apart. I dig a hole for each, and throw a dollop of compost in, mixing it with the native soil before planting. Then it's just watering and weeding until the distinctive nightshade flowers die away and the fruit begins to show itself. Healthy plants will reward you with pounds and pounds of delightful, glossy, melt-in-your mouth eggplant without a hint of bitterness. My favourite varieties are Tango, Orient Charm, Dusky, and Zebra.

I used to make soil blocks for most starts, but as my needs grew, I found that quite laborious. Now most seeds get scattered rather thickly in seeding trays, and I save the block-making for the plants I like to coddle, like eggplant, peppers, broccoli, cauliflower, and cabbage. Since you can fit only fifty two-inch blocks in a tray, they can take up a lot of space on your tables. Scattering a few hundred seeds in one tray helps you capitalize on greenhouse space. I do this for onions, leeks, parsley, chard, and kale. None of these seedlings need potting up, so when the time comes to plant them out, I first move the trays to tables outside the greenhouse to harden off the plants for a few days. "Hardening off" takes a minimum of two days, preferably in at least partial shade and protected from strong wind. The plants are delicate after the humid, protected greenhouse environment and need to get used to breezes and the harsh rays of the sun. Your plants will suffer from sunburn and may even die if they are planted out in sunny weather without the benefit of hardening off. Wind can also be devastating to seedlings that are planted out fresh from the greenhouse. Cloudy, calm weather is perfect for transplanting.

In April, basil seeds are started thickly in trays. Tomatoes, eggplant, and peppers get transplanted into four-inch pots when they have two sets of true leaves. Around this time the squash, cucumbers, and melons are started, also in three- to four-inch pots, and the greenhouse gets pretty full. I use pots for these plants because they have larger seeds and each likes to develop a decent root system. If they are started in anything smaller, they use up the space and very quickly become rootbound. Sometimes I use shelves underneath the tables to make room for more trays but if the trays are placed directly on the ground, slugs, woodbugs, and other seedling-chomping critters are attracted to them and you may find a nasty surprise when you check on your plants in the morning.

By early or mid-May, nearly all of these plants have been cycled through the greenhouse and hardening-off process, and are being planted out in their final spot in the garden. The tables are removed from the greenhouse and I plant tomatoes, melons, and/or peppers in the ground there. It is now warm enough that any seeds that are started in trays after this point can be outside. Although it can be done earlier, I start my cole crops, such as broccoli, cauliflower, and cabbage, in May, due to the greater success I've had at that time. Early cole crops tend to get ravaged by slugs and other pests, and are also famous for bolting before the plants are ready, resulting in small heads. Toward late summer my tables are lined with trays again, this time with winter seedlings such as choy, lettuce, onions, Napa cabbage, and sprouting broccoli.

These squash and pepper seedlings are some of hundreds of seedlings that line our propagation greenhouses in the spring.

Growing Bountiful Broccoli— Rachel

Broccoli is one of those staple vegetables that every-one loves to eat but is moderately difficult to grow here on the coast due to acidic soil and fungal conditions. Broccoli is also a great delicacy for slugs, which make short work of seedlings in the spring.

For reliability, I start seedlings in June for harvest from early August till December. At this time of year, they can germinate outside in the open air, one seed per soil block or plug. When the seedlings have two to three pairs of true leaves, they are ready to plant out. It is crucial that they do not get rootbound in their tray or they will head up prematurely and never reach their full potential. Broccoli prefers a pH of 6.4 to 7, so I add Dolopril (prilled limestone) to the bed to bring up the pH. This helps prevent clubroot, a devastating fungal disease that causes deformed roots and stunted growth. It is very difficult to get rid of once it has taken hold on your land. The broccoli seedlings should be planted a minimum of two feet apart if you want big heads. The plants can get large, and need a fair bit of root room to do so. I dig holes and add a shovelful of compost and a handful of Dolopril, and mix that with the native soil before planting. Seedlings should be planted as deep as their first true leaves, thus burying the stem, which is vulnerable to wind. Water and weed consistently and well, and the plants will grow to large proportions before beginning to head up.

Once the main head is cut, it's not over by a long shot. If you choose a variety that has a lot of side-shoot production, you'll be blessed with broccoli for many weeks. Larger plants translate into larger heads, larger side shoots, and longer production. My favourite variety is Everest. This is a hybrid variety where the heads come all at once. Home gardeners may prefer an open-pollinated variety that ripens one at a time.

I'm fond of direct-seeding just about every-thing else, such as beets, carrots, spinach, braising greens, collards, salad greens, radishes, beans, peas, turnips, et cetera. This is my preference but it is by no means the only way. Other farmers start most crops in trays first, to get a jump on the weeds, for example. I have found that if I can prevent the weeds from going to seed on my farm, I have a lot more success with direct-seeding. I tend to seed slightly more thickly than recommended to ensure good germination in a bed. With the thicker seeding, I am learning to be ruthless about thinning out the weaker or crowded seedlings, because the remaining crop grows to the perfect size and shape, and more importantly, everything is around the same size when it is mature. This aids harvesting tremendously, since you can simply clean out the bed from one end to the other without wasting time hunting for big ones amongst all the crowded seedlings. Thinning is important! It has taken me ever so long to learn this, but it is one of the best techniques for this type of intensive growing-and-selling operation.

The Bottom Line

The burning question: Can you really make a living at organic farming? If so, how much? *Yes, a modest living*, I usually reply. You have to work your butt off for at least seven months of the year, and it takes a minimum of three or four years of experience and a reasonable amount of financial investment to get to a comfortable place. My own path was extremely slow and gradual. At first, this was due to my attitude, since I was in it not for the money, but for the lifestyle. Later, when I began to view the farm as my business, I also began to

have children, and for a couple of years, I experi-enced uncertainty about my land lease. I feel that the financial picture really started to show some promise when we moved to our current farm. In the fourteenth year since my apprenticeship, my farm grossed $60,000 with an acre and a half under cultivation. Given the right circumstances, marketing strategy, organizational skills, prudent investment, and hard work, this could be achieved much earlier. For example, there is a couple in our community who did significantly better than this in their fifth year.

Now on to the real goods: the net total. My non-labour expenses are usually around $10,000 per year. A third of this is for a type of commer-cial compost we use in this area and for other soil amendments. Fuel for the truck and rototiller, truck insurance, hardware, and seeds are also large expenses. In the past couple of years I have hired people twice a week to help on harvest days, and on occasional other days when I am feeling swamped. This usually comes to around $5,000. This past year, I also hired two people for four days a week each from April or May till the end of October at a rate of $10.50 per hour. My total labour expenses were $18,000. In the end, the farm netted $30,000, about $5,000 more than the previous year, though I spent a lot more on labour. This net increase is partly due to having more land under cultivation, but just as importantly, each year brings increased knowledge and more efficient methods, and, of course, the value of the streamlined marketing of Saanich Organics cannot be overstated. Even small amounts of crops get sold, because it takes little additional effort at the sales end. (More on this in the Saanich Organics chapter.) Before our

current marketing scheme came to be, I'd shrug off the small amounts because if they didn't sell at the market or to my one restaurant customer, it wasn't worth taking time out from growing to try to sell them. Now, all those little sales add up.

Compared with conventional agriculture, I can pay a decent wage (with the additional benefit of all the produce the workers want to take home). Hiring people also allows me to enjoy the additional benefits of having knowledgeable people present throughout the season to help with the volunteers who come and go. I was able to step back from the hardest jobs as my second pregnancy progressed, to step away completely for a month when my baby was born, and to spend valuable time with my older son, confident in the knowledge that the farm was in good hands.

If you are able to spend all your time on the land, or if you work as a couple or group, or if you rely on apprentices or volunteers, naturally your labour costs will be diminished. The figures I've shared are not the best around, but they are respectable, and I haven't felt that I've maxed out my gross income potential yet.

Setting Financial Goals

In my first year I made $5,000. Every year since then, my income has increased. At some point early on, I thought that grossing $20,000 would be the cat's meow. Within a year or two of setting my sights on that, I achieved it. For three or four years, I didn't set a new goal. Interestingly, my income increased by only a little. In time, my situation changed; I gave up my other part-time job and became more committed to the farm for my income. Robin, Heather, and I purchased Saanich Organics and started to work together. Our marketing became increasingly organized. When I realized that I had stopped setting goals, I decided to try for $50,000. It took two years to achieve that but my income jumped significantly the first year because I was aiming so high. The next year my goal was $60,000, and interestingly, I made pretty well exactly that.

What if it had been $10,000 higher? My current goal is $80,000.

If you have a goal, your efforts will be shaped by it. Be sure your goal is a positive, definite one. A negative statement such as "I don't want to do as poorly as last year" will not get you anywhere. "I want to make more money this year" is better, but still too wishy-washy. If you really want to make money, be specific. Decide on a figure that will challenge you without being unrealistic. Then pay attention to what makes money for you and what doesn't. Would there be a market for that vegetable or berry if you grew more? Pay attention to what customers are looking for at the farmers' market and grow more of it the following year. Work out how much money each crop makes per square foot (this is described in Heather's chapter; see also Robin's Crop Value Charts in Appendix B). Having a goal only works well if your actions are defined by it.

Our community has changed a lot since I started farming. Back then, very few organic farmers were grossing over $50,000. We've all become better at what we do, and a certain healthy competition urges us forward. The demand for our wares has grown steadily due to health, environmental, and food-security concerns. Even mainstream people know what organic produce means, and many of our customers really have a sense of what it takes to make a living at this type of work. There are more people buying local and organic food, which helps to spawn more farmers. The more farmers who are successful, the more organic food is available, the more people enjoy the food, the greater the demand. Now there are many farmers grossing over $50,000 in our area, and $100,000 is the new benchmark. It is easier for a new farmer to make money at a faster rate, because there is so much more expertise to draw on. My goals were influenced early on by what I perceived to be possible. I remember several of us talking about an article that claimed you could make $25,000 on an acre. That seemed outrageous at the time. It must all be in strawberries, we thought. Now I see apprentices move on to start their own farms and they do as well in their first year as I did in my fifth.

The Human Landscape

One of the best and most extraordinary things about organic farming is the quality of the people who are attracted to this way of life. Our immediate community of Saanich Organics, and the larger organic-farming community, includes some of the best people I've ever known. We are environmentally, socially, and politically conscious. We are respectful, supportive, have a sense of humour, and work together as well as we can. We are bonded, due to both the joys and the trials of farming life. We try to live in a manner that is aware of our ecological footprint, and, while being realistic about the importance of earning a living, seek to deepen our quality of life rather than to increase our quantity of material possessions.

I have come to depend greatly on Worldwide Opportunities on Organic Farms (WWOOF). Participating countries have their own registry of organic farmers who are available to host travellers seeking an integrated working experience on a farm. Visitors stay for varying periods of time, helping out in exchange for room and board. Though most of the people we have hosted have had no previous

Elias ready for some harvesting action with his wheelbarrow.

Elias Harvesting—Rachel

The year my son, Elias, was three, I started to enjoy his presence with me on the farm more than ever. He often had friends over and while I was working, I'd hear the sounds of pattering feet and screams and cries of "Let's go pick some *stwabewwies*!" and not long after, "Let's go pick some *cawwots*!" He explained to his friends in an authoritative manner how to pull them out without breaking the tops, just as I had explained to him. After the *cawwots*, it's cherry tomatoes. The best day was one Monday in July when we were harvesting for the box program and restaurants. I was harvesting broccoli, and an employee was harvesting chard and kale bunches. I looked over toward the cucumber rows and there was Elias, quietly harvesting cucumbers off the vines into a bin, just like the rest of us!

experience of this nature, they have always brought great enthusiasm and positivity to the work, and that more than makes up for lack of knowledge. I look forward to their arrival for the extra help, but I also know that my family and I will be enriched by the experience. I've enjoyed immensely the formation of new friendships, many temporary, some lasting, and watched fondly as my kids warm to them, start to call them by name, follow them around, and create games with them.

Normally boring jobs such as weeding can be transformed by exchanging personal stories rooted in our various cultures, family experiences, and social fabrics. WWOOFers bring fresh eyes, thoughtful questions, new energy, and new ideas to systems that function pretty well but can always be improved. I try to be open to the skills and qualities embodied by each individual, and to be aware that certain tasks can bring out those qualities and deepen his or her experience.

Some WWOOFers leave their mark on the physical landscape. Jesse arrived from Maine one August and stayed for two weeks. He was trained as a stoneworker and was skilled in carpentry. When he departed, we had an elegant new firepit beneath one of our graceful oak trees. It consists of large, carefully placed granite stones gathered from around the property and is surrounded by benches of flat rocks supported by smaller rocks. The stonework is simple but consciously done, and it blends well with the landscape. On a more practical level, Jesse also built a frame and countertop for the bathtub we use for washing produce, thus finally completing a job that had been on my mind for months. He returned the following year and built another bathtub frame, a shed, a composting toilet, and a treehouse for the kids.

These wwoofers were already experienced growers and harvested these potatoes in record time.

Some leave their mark in less tangible ways. Jamie from Santa Cruz formed lasting friendships with everyone around her, and blessed us all with her generous spirit, genuine caring for others, and awareness of the foibles of the human heart. Nadine from Germany brought her positive spirit, love for children, and openness to everything new. Some contributed their absolute willingness to help, learn, and fully experience life on an organic farm. I connected right away with Jenn, due to her down-to-earth personality and witty insights into the human condition. Max and Claudia from Germany shared their intellect and insatiable curiosity. Anastasia offered her open heart and love for family, and infected us all with her eagerness to embrace life

with every ounce of her being. Another Jamie also connected with everyone around, with his thoughtful wit, enthusiasm, and willingness to share from the heart. We were blessed with music by Kim and Sarah, who gave us a tiny personal concert on our balcony the night they left. Roslyn from Northern Ireland was so quiet at first I thought we'd never get to know her, but as time passed, we enjoyed her infectious laugh and her travel stories. Tracy from Nova Scotia brought her upbeat and easygoing personality.

Jess and Kat were wise beyond their years and approached everything on the farm with enthusiasm, and every person with warmth. Dara and Kate from Idaho and Northern California brought their different experiences and intelligence to many late-night discussions. Izad from Malaysia shared his interest in personal development, and cooked several memorable meals. Sarah and Chris brought a lively energy, knowledge of communal living, and previous farming and carpentry experience. Drew returned a second time and deepened her interest in food growing and sustainable living. David and Gemma from Barcelona told stories of their two-year trip around the world and gave impromptu Spanish lessons. Aimee and Tim from New Zealand came with their nine-year-old son, Oliva, who bonded instantly with my son. Every person who has passed through our lives and has shovelled compost, sowed seeds, weeded, harvested, cooked, taken endless photos of the rainbow chard, or read books to our children has left their mark. For them we are ever grateful.

Apprentices and employees have in more recent years contributed a huge amount to the success of the farm. They come to learn, but I always feel that I learn at least as much from them as they do from us. They leave the memories of their own special interests, crops grown and favourite or least favourite jobs, and over time, aspects of their personalities seem to linger and become part of the history of the land. Catherine loved working with the tomatoes, and was always willing to be the one sweltering in the greenhouse on harvest days. She kept coming right through the winter, bringing her positive energy to some trying times. Kat had a deep interest in growing food and, with the opportunity to learn and grow, has become an accomplished farmer herself. She now co-manages the farm with me, and has brought it to new heights.

Peter and Randi apprenticed for a whole season and made the most of everything, setting up an outdoor kitchen and building a beautiful cob oven out of which the delightful aroma of Randi's baking drifted on many an occasion. Peter would be out in the garden before anyone else, watering and doing tasks without being asked. Carla and Justin jumped into the work with the gusto of passionate farmers, taking care of the chickens after hours and tending their own ever-growing container garden. Karlyn set the pace from the beginning, working as fast as any seasoned farmer while keeping us entertained with hilarious stories. Both laughter and tears flow easily in such a natural setting, where people are real and the defences of normal society are down. I have seen many apprentices come and go from our farms in the last fourteen years, and marvelled at the quality of people joining our community. Some have moved on to other things, and a few have stayed and begun farms of their own, enriching our lives and our community in the process.

The Three Oaks crew from 2010: Peter, Kat, Carla, and Randi.

Scything Bliss—Robin

Managing the grass that grows in the paths and buffer zones between the fields was always a noisy and "gassy" chore with the Weed Eater. It was a necessary evil. Rachel describes "being able to think clearly again" once the encroaching grass is beaten back and the paths are levelled. It always feels counterproductive when you think about the other nine-tenths of the world where people walk miles for fodder, and we see the lush grass as a nuisance. I tried, with little success (very little) to raise lambs on the grass in the buffer areas (see the Lambs section), but it was just too much work. Then along came Fairlight Vido.

Fairlight is from a scything family in New Brunswick, and she came with her scythe to Rachel's farm. It was a real Austrian scythe, with a light blade and a snath (handle) handmade from a tree branch. The only word for a sharp scythe in a stand of tall grass is "bliss." The crisp cut, the meditative *swish swish* sound that blends with the birdsong and the wind in the trees, and the empowerment of mastering a traditional tool that works better than a power tool, are all part of the experience. The Vido family scythes grass only in the early mornings when the water content is highest, so the grass cuts crisply and cleanly.

It's essential to purchase the right kind of scythe, fitted for your hip height and arm length, with a lightweight snath and blade. There are dozens of types of blades intended for specific uses. Technique is key for both cutting and sharpening. Your stance must be balanced so that the blade stays level and is riding on the ground for the entire semi-circular swing. Allowing the blade to rest on the ground means you use very little energy. This takes practice, because the temptation is to lift it up. But it should feel almost effortless. Each swing should end at the same place so that when you are

Kira learning to use the scythe during a monthly apprentice enrichment workshop.

done, there is a windrow of grass on your left side that can be efficiently collected with a pitchfork. Ideally, the ground should be free of bumps, rocks, and obstacles.

Then comes the sharpening part. When I first heard that you had to sharpen the blade every five minutes or so, I thought I could never be bothered, but it's fast and it makes a world of difference, so I don't mind. Scythes come with a long, thin sharpening stone and a holster that clips onto your pants. When you feel the blade isn't cutting as cleanly as it could, stand the scythe on its handle and file the blade with quick strokes. There is a technique for sharpening; a rim on the blade helps beginners angle the stone so it's straightforward and easy to get a perfect edge. We continue to learn about scything techniques. See www.scytheworks.ca and www.scytheconnection.com for more inspiration.

If I could distill the essence of what this life of growing food is all about into one word, that word would be "promise." The promise of the beauty and bounty the next year could bring keeps the dream at the forefront through the wet, cold winter. Come spring, the promise is experienced many times over, from seeding, watering, and planting out seedlings, to witnessing the flower giving way to the stub of what will become that glossy purple eggplant, or sweet succulent pea. It's the promise of a type of heirloom seed sourced from a nearby farmer as dedicated to ethical agriculture as you are. With the longer-term crops like asparagus and fruit trees, which will take a few years to fulfill their promise, the dream remains in the air like a kiss blown on the wind. It's the promise you experience walking down farmers' row at the market, complimenting others on their lovely produce, secretly hoping that yours will be as stunning next year.

In a larger sense it's the promise of affecting the reality of food production, first in the immediate area, then farther afield. As communities like ours continue to grow and thrive in other areas, it's my hope that the modern industrial agricultural system that impoverishes farmers, their livestock, and their land will give way to the promise of an agriculture that is truly sustainable, one that reflects respect for the land, the people who eat that food, and the people who grow it.

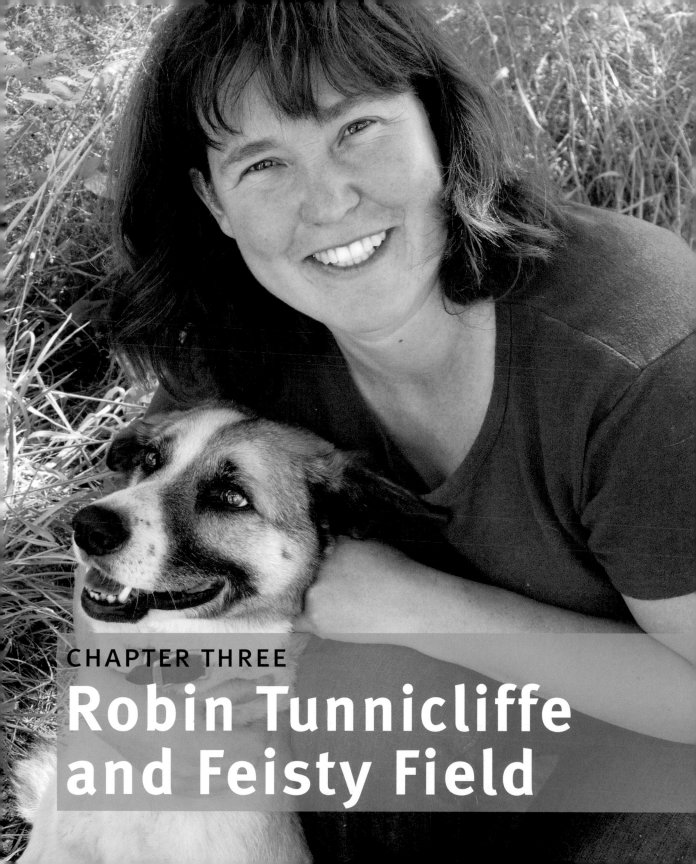

CHAPTER THREE

Robin Tunnicliffe
and Feisty Field

Feisty Field with the many salad beds covered in floating row cover to prevent flea beetle damage.

A question I get asked a lot is, "How did you get into farming?" I sometimes wonder myself. I know that I spent long hours in high school daydreaming about communal living. I sought out books on alternative living at every opportunity and I was attracted to agricultural courses at university. Something resonated within me about the rightness of being self-sufficient, which was contrary to the problems of overconsumption, corporatization, and being out of balance with nature. However, I think it was the time I spent travelling and working on farms in South America that really cemented the notion. I was heartened by the families who were working together on the land. I loved the tranquility of their lives, the fun of community, the practicality of growing one's own food, and the age-old traditions of the farmers' market. While small-scale farming worked well in the South American setting, the challenge was to take this model home with me. Fortunately, it already existed.

I attended a farming conference in 1998, shortly after arriving back in Canada, and I learned of an organic-farming community on the southern tip of Vancouver Island. I was offered an opportunity to apprentice, and I accepted wholeheartedly. During the apprenticeship, I learned farming and marketing skills but I also participated in the farm community, helping to organize a barn dance and attending farmers' meetings, work parties, and potlucks. I found the company of like-minded people inspiring and very welcoming. When the season on the farm ended, my mentor helped me find land to lease. Ten years later, I am still on that same piece of land. Every year, the farm becomes more productive, more profitable, and more fun. Challenges still side-swipe me when I least expect them, but that's what farming is all about.

The farm has grown to a full acre under cultivation: three-quarters of an acre at the original Prospect Lake site, now affectionately referred to as Feisty Field, and a quarter-acre at Northbrook Farm. I have a very secure arrangement with a long-term lease. I added the plot at Northbrook when Andrew, my last partner, also started farming full-time because we were concerned about making enough money to support us both. What happened when we started working together was that we became more than the sum of our two parts: having Andrew on board more than doubled production. Now that I'm back farming on my own, I'm certain that output is a function more of labour than it is of land. An acre is plenty of work and provides enough income for me to live simply. I have no plans to expand, although the dream of land ownership, or at least a life-lease arrangement, remains alive in my heart.

Apprenticing

I can't say enough in favour of this type of hands-on learning. When you're taking on a practical occupation like farming, the benefit of experience makes all the difference. Why make your own mistakes when you can learn from someone else's? Apprenticing can take many forms: full-time or part-time, over a season or for very short periods. I lived on Tina Baynes's farm for about four months and worked full-time in exchange for learning, room, and board. I loved every minute of it. It is a truly blissful way to farm: having all of the experiences and none of the worries. Combined with great, home-grown, organic meals, it makes for a pretty nice summer. Of course, it's not always possible to take a block of time off to learn farming, but it may be possible to work out a schedule with a nearby farmer to go there once or twice a week.

A farmer once told me to apprentice at a farm that makes money. I thought this was a bit shallow at first but now I see what he meant. It definitely doesn't mean choosing farms that are flashy with new equipment and well-dressed farmers; it means choosing a farm that is working successfully. Farms that make money are running purposefully, and the skills you learn from those farmers will translate into a career for you. There are always apprenticing horror stories, like slaving away for some rich folks at a hobby farm, or ending up helping to fight someone's losing effort at farming. Before committing to anything, go and visit the farm, and possibly work for a few days. Ask where they are selling their food and how regularly. This should give you a sense of what they are about. Remember that you are not a slave, and you should be treated as a welcomed guest, if not part of the family. Your labour is fair trade for any questions you have. You should be able to try out your suggestions about how to proceed with certain jobs. Farmers often give apprentices more help after they leave the land. They will sometimes visit your new site and give

you suggestions; they might even help you market, or problem-solve when you come to a roadblock. A good apprenticeship is a lifelong relationship.

Starting the Farm

Starting my own farm was a very exciting time. There were lots of highs and lows, but it was always challenging and intensely real. Looking back on that time now, as a hardened veteran, I see how very insecure I was: I worked so hard, and I was so proud. The smallest threats to my success were crushing and tears came easily. I felt tested in many ways: endurance, creativity, problem-solving, budgeting, and more. I also felt very much on my own as I embarked on this adventure, against the counsel of many people I respected. At the same time I felt a network of neighbours, new friends, and colleagues, who really wanted to see me succeed, crystallizing around me. They were cheering on the sidelines and ready to help when they were called upon. Although I can recall wiping away tears with muddy gloves, I look back on that time fondly, and I see it as a very formative experience and a time of great confidence-building. I think three elements have to come together to make a farm work: soil-building, personal farming style, and business skills. For me, the incubation period for these fundamentals was about three years.

Building the Soil

Getting the soil primed to grow annual crops was my first task. This involved a delicate balancing of the main soil elements: humus, pH, and fertility. I had a soil test done right away. Most agricultural-supply stores will take your samples and help you to interpret the test results. I paid special attention

to building the soil, especially in the hard, rough spots, to ensure good, even production. I started out with a pretty balanced soil test, but it still seemed like I was just pouring my money and sweat into the ground as I dumped bag after bag of potash, rock phosphate, and lime on the land. All that compost that I hauled out into the field just disappeared! It was truly fortunate that I believed in the process, because it sure didn't seem like it was going to make a difference.

Then, in the spring of year four, I realized that I wasn't having to smash up the clods of earth with the back of my shovel to get enough loose dirt to cover my transplants. The weeds were actually being pulled up with their roots intact. I was getting big spinach leaves consistently down the rows. The soil was loose and loamy in my hands. I dug my hands in deep, and brought the beautiful, dark, rich earth to my face and breathed deeply: the sweet smell of success!

Hoeing Techniques—Robin

I once held a farmhand job where I sometimes hoed all day long. Now I always look for light-weight, long-handled tools. It's always better for your body in the long run to do lighter tasks more often than a few huge, back-breaking loads. A smaller tool, sharpened often, may be the way to save your body. A long enough handle will help you stand straighter and prevent back-ache. Hoes should be held like a broom, with the top hand placed palm up underneath the handle. The instant you put your hand on top, palm down, your back is bent. After an hour of this, you'll be ready for the geriatric home. Don't compromise your posture. Learn to hoe standing up straight even if it feels strange at first.

A cross-section of the field showing drip-irrigation lines and the diversity of plant families side by side.

Developing a Style

The second element that had to incubate was my relationship to the land and how I work as a farmer. Over the years, I have visited and talked with many farmers, and I've come to the realization that each farm is a unique combination of personality and physical elements. A farmer's individuality really shines through his or her operation, and when it's a good fit, it's a beautiful thing. The challenge for me was to figure out what I liked growing and what grew well for me. I experimented with veggies, chickens, and livestock. I settled on mostly vegetables and strawberries but I tried many varieties of many different veggies, and this learning process continues. I also developed a style of how I do things—systems for watering, harvesting, et cetera—and favourite tools that help me do my job. This all takes time, and farming gets easier over the years.

Markets can always be built, so make sure to try crops that are easy and successful for you. Is your site windy, hot, cold, wet, dry, rocky, shady? There are ways to use all of these conditions to your advantage, or at least there are ways not to be constantly fighting what you can't change. Windy sites may have fewer pests, so you might be able to grow crops that are susceptible in other areas. Hot sites can yield sweeter tomatoes; cold sites may be more productive for broccoli and cole crops. Don't look at extremes as handicaps, but rather use them to your advantage. This is the whole key to marketing: what can you grow that others can't?

Test the boundaries of your plot and get to know your land. What planting dates are reliable for your site? While local farmers might tell you to plant squash out on May 24th, you might have a sheltered microclimate that gives you a week's advantage. Only experience can tell. Do you like growing in the winter? If you can handle the cold, and have a suitable climate, you may want to extend your season. All this getting to know your land and yourself is a lifelong learning process, but in year four, I found myself with more confidence ordering seeds, contracting to grow in quantity, and planning for continual supply at the market. Obviously, I still had crop failures and surprises, but I had more understanding of cause and effect.

Business Skills

The third element is business skills. Never forget that your farm is just as much a business as any other commercial enterprise. Building up the reputation of your business and establishing yourself as a reliable supplier takes time. If you haven't run a business before, you might want to consider taking a business course. I ended up taking one as part of a community skills-training program and I was surprised at how much was relevant.

I was very hesitant to write a business plan, but I did it as an attempt to secure investment in my farm. This failed, because my bank didn't recognize organic farming as a viable business and I didn't have any collateral. So I threw my business plan in the recycle bin and moved on. Now I have a different view of business plans and their role for the beginning farmer. The key is to write the business plan for *yourself*. It is a great way to harness your creative energy in the planning stages, and, very importantly, it helps you explain your ideas to others. This is very necessary if you are leasing land from a land trust or from landlords who are not familiar with sustainable agriculture. It can also help you communicate with partners, family, and neighbours so they can understand what you are proposing and see where you need help. Don't be afraid to leave blank spaces or write in pencil. Just be honest and open, and the plan can really help you define your business in your own terms and problem-solve into the future.

There are many styles of running a business, and you have to decide what works. In the beginning, I was convinced that I was going to do all direct marketing, to maximize my profit. I did farmgate sales, a box program, and farmers' markets, plus sold to grocery stores, and a restaurant. I called everyone myself and delivered to each customer. I enjoyed meeting all my customers and I found it very satisfying to get their feedback. It was a great learning experience to get to know what people wanted, and how they wanted it packaged. It was only after

Business Plans

Think of business planning as a way to harness and focus your creative energy during the planning stages of your farm business. Launching a new business can be stressful; this exercise can help you get a handle on the unknown elements and give you a way to proceed with your research. A completed business plan is a tool with which to share your ideas and vision with others. Often, friends or family can be a source of funding for new farmers who don't have any collateral for a bank loan. Showing commitment to business planning is a good way to demonstrate your serious intentions. See the Community Farms Program website for a good business plan template at communityfarms.ca.

I started wholesaling to Saanich Organics that I realized how much time I was spending on the marketing end of things, and how that translated into lost production on the farm. This is an equation that only you can figure out, and it will change over time. It is important to constantly reassess what your time is worth, and which parts of the business you enjoy.

With all businesses comes recordkeeping. This is an area that I desperately avoided in my first years of farming. I wasn't making any money, and I really didn't want to see exactly how much I wasn't making written down on paper. My other jobs funded the farm, and I didn't want to think very much about it. This was my big problem. It is amazing the clarity that comes when you tally up everything at the end of the month, or as I do, at the end of the year. If paperwork isn't your forte, it might be worthwhile hiring a bookkeeper to make sure that you are on top of your finances.

You get to see very clearly which crops are worthwhile and which are not. I like to calculate the return of each hundred-foot bed. For me, a bed can't make less than $600, and preferably $800, in a season. Some people think about it in terms of square footage. I farm more intensively than most. Last year, in addition to Andrew's and my full-time labour, we hired a full-time student from May to September. That meant there were three of us working an acre. Again, I believe income on an organic farm is related more to labour than it is to land area. When your beds are nicely balanced with fertility, seeded on schedule, well germinated, thinned and weeded at the right time, and harvested promptly, the sky is the limit. When you write it all down, you can see how to hone your crops until you are satisfied with your income. But be aware that things do look better on paper.

Nadia, Grace, and Robin at an October farmers' market. Nadia was a homeschooler who spent a few days a week helping at Feisty Field over the years. Now she is a busy 17-year-old and does paid work at the farm when she can.

Of course, it isn't all about money. Some crops, like eggplant, are just aesthetically nice to have, while others, like purple sprouting broccoli, are a drawing card for customers who will then buy an assortment. However, it is not wrong to want to make money. A farmer works really hard. Why shouldn't he or she have a good standard of living? I was just scraping by for a number of years because I somehow believed that I *should* be poor, that by choosing a farming lifestyle, I was choosing poverty. It wasn't until I decided that I needed money or I was going to crack that it started to change. Attitude has a lot to do with the flow of your business. Make sure that you are being fair to yourself and to others, and your business will flourish.

I think all farmers face ethical questions when it comes to the prices they charge. Food is different from other commodities. It's really hard to think about the beautiful, nutritious food that I grow and to know that it is priced out of reach for poor people. I wish that everyone could experience the vitality of locally grown food. Many people choose not to; they will spend their money on other luxury items and then complain about the cost of my produce. I don't worry about them, but I really feel that people need to value food more and prioritize differently, because cheap food from afar has severe consequences for society and the environment. I do worry about the people who genuinely don't have access to good food. As I've pondered this situation, I have come to the conclusion that it is really a social problem, and more than any one farmer can shoulder. As a farmer, I have a clear conscience about how hard I work, and what I need as a person to sustain myself financially. I think we need to continue to work for justice in the food system, but we must also earn a fair living so we can continue that work.

The Physical Elements

Choosing your Site

The more thought you can give to your site, the more it will pay off down the road. Because I was leasing, I had more options than someone who already owns land: I could take my pick between several plots. In retrospect, I wouldn't have chosen the piece on which I now work, because it is very low-lying. We are the first in the community of farmers to get a frost and the last on the land in the spring because it takes so long for the soil

to drain. We have good sunlight in the summer, with only the top section of the field shaded late in the day, but in the winter, my sunlight is really limited by trees, and this slows the growth of the greenhouse crops. Had I chosen a south-facing slope, I would have more options for extending my season and fewer trenches to dig for drainage.

If you are looking to lease, it may be worth your while to study a map of the area that interests you. Take a drive down the rural roads with the prime farmland, and put flyers in mailboxes, describing who you are and what you'd like to do. You never know who might call, and it can't hurt to have many options.

Some aspects of my site are so ideal that they almost make up for my less-than-perfect locale. For one, I have great landlords, who understand the long-term nature of my project and are supportive in helping with infrastructure and emergencies.

Another great aspect is the close community surrounding the farm. Quite a lot of traffic must pass by on the way to the main road and as a result, many people have observed the progress and feel a bond to the farm. They really cherish the farm as a community resource, and I get a lot of friendly waves, volunteers, and enthusiastic customers.

Before Signing a Lease

Clear communication is key to a good lease arrangement. It is essential that you verbally paint a picture for your prospective landlords about your plans for the land. I was able to present my landlords with a business plan that I had written up. While I felt a bit nervous because I had no idea if the proposal I was presenting was even possible, the landlords got the impression

Garlic is cleaned at harvest because the outer skins peel off easily when first pulled. It is then hung for a month on the side of the barn where it has good airflow but protection from the rain and dew.

that I was very serious about this endeavour. Even more important, I was able to convey the amount of labour I was about to invest, in anticipation of long-term gain. If landlords can't guarantee you at least five years, you'll have to think very seriously about whether or not it is worth your while. One of the big problems for tenant farmers is that our society is very transient and unstable. Many people who think they are settled will get a better offer from elsewhere so they up and move. Try to feel for groundedness. (See Appendix B: Robin's Lease Agreement)

I believe that one of the most essential skills for a tenant farmer is diplomacy. In my situation, the family lives on the land and must pass by the field several times a day. In other situations, the farmer may rarely see the landlord. I can't say enough about keeping communication open and friendly. Make it easy for them to plan around you by giving them all the information they could possibly need. Share your long-term plans so there aren't any surprises.

Don't forget that there can be advantages for the landlords. One main reason people lease out land is for tax purposes: if they are savvy about taxes, they can save a lot of money. You should educate them about that if they don't know, because all the advantages you can bring to them will only help the relationship. To find out all the possible benefits for landlords in your area, talk to other farmers and agricultural organizations.

An issue that can wreak havoc on tenant farmers is water. One of our friends started out on a promising bit of land, but was soon held hostage by a water shortage: she had to share the limited water with three households, and after a season of water-deprived crops, she was given her notice. Ask about water, about the possibilities of connecting to city water if a well dries up, and if the landlords have any plans for emergency backup. Ask about traffic on the land, which can be a big bone of contention. You will want to have work parties, perhaps host a few garden tours, and possibly even have a farm stand. Discuss this and create a working vision. It's really important to understand that there is a huge reality gap between the public perception of farming and the real thing.

Be sober and realistic when entering a lease agreement. It's tempting to be romantic and idealistic, but think of the everyday farm operation and don't make too many concessions. You will be spending many hours on this property and it needs to work for you.

Community Farms Program

The Land Conservancy and Farm Folk City Folk, two BC organizations with strong agriculture advocacy backgrounds, have teamed up to help new farmers get onto farmland. The Community Farms Program is focused on reshaping policy to make it possible for many farmers to live and work on a single piece of land. The organizations are pushing the idea that the future of farming looks different: Farming isn't always going to take the form of a husband and wife living in a single dwelling on a piece of land. Organic farming is labour-intensive, and can be a low-return occupation for many years before the land is in shape for high production. This requires the pooling of labour and resources. Anyone who is interested in growing food for the community should be able to start up, and the Community Farms Program is about teaming up, sharing resources, and growing food.

Site Design

I was fortunate that I picked out my site in November and got to roam around the land almost daily through the winter. Putting a lot of thought into how you lay out your farm will benefit you through the years. There are many books about design, and they will encourage you to think more thoroughly. Permaculture is an inspiring philosophy of ecological landscape design with a focus on food production. It is worthwhile taking a look at any of Bill Mollison's books. I found *Permaculture: A Designers' Manual* a bit heavy going, although fascinating and very relevant. His lighter, *Introduction to Permaculture*, on the other hand, was so gripping that I could barely put it down. While the concepts of permaculture are wonderfully thoughtful, they are hard to incorporate into a high-output, annual-cropping system like a commercial farm. Much as I wanted a permaculture farm, I realized that I wanted to farm for income, and I had to make money first.

Always imagine your enterprise growing bigger than you currently think is possible, and design for it. For example, I fully envisioned myself as a seasoned farmer, wandering the field, harvesting this and that into my basket, and turning over the soil with my shovel. The reality, five years later, is that I need two pickup trucks to haul the produce to market. I hire a tractor twice a year to cultivate different areas, but I am limited in the size of tractor that can get into the field because of my nine-foot-wide gate. A standard twelve-foot gate would have been a better idea. Thank goodness I listened to my landlord and left a truck path in the middle of the field, and factored in access for his tractor. I really did not want to—I was almost religious about not having machinery on the land—but I've changed, and I'm thankful that my site design was flexible enough to adapt. I know the only thing that we can count on in life is change, so plan for it.

Another important concept in design is how you plan to water. Water is crucial, and you need to think about the source of your water in relation to everything else. Think about the water pressure waning over distance and about whether you'll be using drip tape or sprinklers to irrigate, and this will determine some geometry and dimensions. For example, drip tape distributes water reliably over no more than a hundred-foot span. This might influence the length of your beds. Some irrigation doesn't work so well with contours, so straight beds will make things easier. When Rachel started out, she had a circular farm that was beautiful in form and function for its size and she watered with a rotating sprinkler that worked perfectly. Her farm design and watering system fit well together.

Think about light, the path of the sun, and how this might change over time because of trees or new buildings. It is a shame to see a greenhouse built in the wrong spot. Think about land-use patterns, and time- and energy-saving methods. If you have animals, make sure they are conveniently located so that feeding and watering them isn't a big trek, especially uphill!

Irrigation

I have tried almost every type of irrigation. I have been known to refer to my watering system as my "irritation" system. In my experience, no type of irrigation, except hand-watering, works perfectly. If you start with that premise, you'll probably be a happier farmer. Better yet, always assume the

Drip irrigation is a very efficient way to deliver water directly to the roots of plants.

system is not working until you're certain it is. The best irrigation system is the farmer's foot-steps. Don't be sold by irrigation dealers claiming perfect distribution; it's just not true. Irrigation is a lot of tweaking and fiddling and experimenting and maintaining.

Andrew gave me another view of irrigation. He loves it. He had a kit with millions of pieces, and whenever a problem presented itself, he loved to assemble all the bits like a big Lego set. He loved to talk to the dealers and find out what new emitters would spit out how much water per minute under what different pressures and sets of conditions. However, he didn't have any illusions about perfect water distribution either.

Let me tell you a tale of woe that will instil in you a fear of automatic irrigation. Rachel, Heather, and I bought a greenhouse together with the hope of having huge returns on a

salad greens operation. We set up an automatic sprinkler system with the perfect heads, on the right diameter of hose, running at the right pressure, set to go on at the optimum intervals. It couldn't fail, right? It became clear that there were dry patches and soaked patches but when we checked, everything seemed to be running fine: the heads were clear, the filters were clean, the pressure was right. We had jumped on the salad greens operation because the irrigation seemed so easy. All of us were working full-time on our own operations so we didn't have time to thoroughly monitor the big greenhouse. As it turned out, when we took off the timer to put it away during the fall clean-up, we realized the valve had only been opening halfway.

That cursed piece of plastic cost us countless hours of worry and frustration, and thousands of dollars in lost production. The moral of the story

Robin transplanting in the early days of the farm.

for me is not to trust technology. Just because it's slick and expensive doesn't mean it works. If you're low on cash and time, stick with simple garden sprinklers for everything except tomatoes, which can't handle overhead water. Sprinklers work well, and I think the produce looks nicer when watered from above. The drawbacks of course are that it takes a lot of time to move the sprinkler around, because you can't properly cover a big area at a time, and the water can take a long time to reach the depth you need. You also waste a great deal of water spreading it where it does not need to go. When you're ready to invest, I would recommend drip tape. It's relatively cheap, and can be assembled with little skill.

Soil

There are advantages and disadvantages to any type of soil, except for the loveliest of loam. The important thing is to recognize your soil type, and plan around it. Sand is harder to keep laden with nutrients because water runs right through it. However, weeding in sandy soil is a breeze. In sandy soil, plan for root crops, which can grow effortlessly and don't need much nutrition. Experiment with greens but know that they have high nutrient demands and may not do as well. Most importantly, plan to get lots of water to your plants, possibly daily in the hot weather. Sandy soil and water shortages don't work together.

Heavy clay like mine can seem like a terrible curse, but as long as you can break it open and get the seedlings' roots covered, they will grow, because clay has inherent richness. Add compost regularly and you'll be well on your way to loam in a few short years. The good news is that clay holds water well, which makes for less summer stress on the plants. The bad news is that you'll be twiddling your thumbs in the springtime, waiting for the soil to dry up while your neighbour with sandy soil is already planting. In clay so heavy that you can't crumble a dry chunk in your hand, contrary to logic, your best bet is transplanted greens. Root crops can't get any size to them under that pressure, but surprisingly, chard, kale, and even lettuce can do well.

I dug a lot of compost deep into the soil in my first year. With this method I had some amazing crops. I also worked wood chips and gypsum into the soil. Wood chips decompose slowly, which helps in the long term. The gypsum, a crushed rock powder, bonds chemically to the clay particles and prevents them from bonding to each other. While I can't say how effective each method was on its own, I am certain the combination of these three things worked, because my heavy clay is now loamy clay.

Weeds

My last partner once tried to persuade me to market my garlic as organic "shade-grown" garlic, because of my jungle of weeds. A visitor to the farm in the early years commended my commitment to the "One Straw Revolution" (based on the book of that name by Masanobu Fukuoka, about his natural-farming methods), because she thought I was trying for the never-weeded look. I failed to tell her that I was actively spending hours each day weeding the field. Visitors to my farm now comment on how there aren't any weeds, and are full of praise for how tidy my field looks. What's the difference? More labour certainly helps but I credit a lot of the change to the excellent condition of the soil.

When I started farming, the clay soil was rock hard, and so crusty that I really wondered how anyone could claim that a hoe was a useful tool for weeding. I literally had to pull weeds by hand, and then they would only break off at the soil's surface, leaving a healthy mass of roots to regrow the following week. I couldn't do anything about emerging weeds because the soil was so hard I couldn't disturb the roots. I tried to mulch using hay, but I just ended up bringing in more weeds. I was pennywise and pound foolish, and I wouldn't use enough to really suppress the weeds. I also tried mulching with composted horse manure, and I imported a grass problem because of the weed seeds in the manure. I remember looking helplessly around the farm in late July, as thousands of thistle seeds floated in the summer breeze, distributing themselves evenly over the farm. Newly seeded crops would be overwhelmed by a flush of weeds, and I would lose the whole bed. I used to cry about the weeds.

Then two things happened simultaneously: Andrew arrived in my life and the soil reached the stage at which it could be hoed. He spent hours the first spring hoeing the field, while I tried to convince him that it was useless. He hoed, and hoed, and hoed. He would show up with yards and yards and yards of fish compost (made from finely ground fish waste and wood chips) that he would use to mulch the newly weeded beds. He didn't care about the initial cost of the compost because he knew from his years of landscaping experience that it would pay off tenfold in a few years. He was right, and I am forever grateful. Andrew's touch definitely helped change the way I farm.

Now weeding is actually fun. Yes, I really did say that. The soil is loose and fluffy from all the compost. Weeds pull up with their roots intact—even dandelions and thistles. I can zip down a bed in fifteen minutes or less; with a gentle scuffing action I can get all the emerging weeds, and then I don't have to do it again because I till or fork the bed over when I harvest. Every year, fewer weeds emerge because almost none go to seed. Smartweed (*Polygonum persicaria*) is hardly seen anymore, and I used to battle whole thickets of it.

The important lesson in all of this is don't despair, and add compost. In your first years on clay soil, you may not be able to get ahead of the weeds. Try, but don't kill yourself, because once your soil has a nice tilth or structure, everything will come together. I suggest that you not open up too much ground in your first year. Paradoxically, you can make more money on less ground because you'll get way, way, higher yields from well-tended crops. A last tip is that if you have a bed that is overgrown with weeds, it's okay to till it under and start again. It actually may be more prudent to do this, because the plants can get stunted from being shaded and strangled, and weeding can be very damaging to them because you'll disturb their roots. It also takes a very long time to clear out all the weeds from an overgrown bed, time that might be spent more profitably elsewhere in the garden, or better yet, in the shade with a cool drink, repeating to yourself, "I am one with the weeds . . ."

Greenhouses

While this may be an odd comparison, greenhouses are quite like livestock in the way they tie you to the land. They need a bit of attention every day: in the late spring and fall you need to get down to the field early to open them up, and you need to time their closing to catch the maximum warmth for the evening. In the summer months you can just leave them open all the time. In the winter, they need venting some days, and of course, they need to be relieved of snow load. If you choose to heat your greenhouse, this will require more daily maintenance.

Less than a foot of snow will damage a greenhouse on the west coast because the snow is quick-falling, cement-like, and likely to be followed by heavy rain. When it is too hard to get the snow off, it important to remember that the plastic is worth much less than the structure; if you have any doubts about being able to remove the snow, slit the plastic to release the pressure on the frame. To deal with a possible snow load on glasshouses where removing snow is awkward, I would recommend a heat source, lit as soon as the snow starts falling. Last year I bought a telescoping pole that extends to ten feet with a broom on the end. This is a great tool for knocking snow off a greenhouse and other farm structures.

It is a big job to build a greenhouse, no matter how simple the kit seems. I have experience with building three types of greenhouses and I will comment on each. But first, a general comment: either buy one that doesn't require levelling or hire machinery to do the levelling. When I started, I had no money and an iron determination that I could do it all, but I know now that all the digging I did has taken its toll on my body. I found it quite intimidating to approach machinery operators to describe a project that I myself was not sure about. The good news is that most machinery operators have done many projects just like the one you may propose and

West coast snow, sometimes referred to as "slement" because of its wet and heavy consistency, is a real threat to greenhouse structures. Karen and Elias use a telescoping pole with a broom on the end to brush off the snow without damaging the plastic.

can possibly give you good advice about how to tackle it. In an extra couple of minutes they might even cut you some drainage ditches or do some other side project that would have taken you a whole day. Ask your neighbours who they use for projects, and call them up.

The Glasshouse

When I first met with my landlords, they mentioned that they had rescued a glasshouse that was going to be demolished. I was ecstatic because I loved the idea of growing under glass. I was less ecstatic when I saw that it was in many bundles and stacks, in about a thousand different pieces.

The first step was to choose a site. I had read that it is best to orient a greenhouse east-west so that you get a good southern exposure. I chose a sunny spot, squared off the corners, and started digging. The next step was to build the cement

foundation on which the glasshouse would be set. I had to dig out a four-foot-wide, three-foot-deep footprint around the fifteen- by twenty-foot area of the greenhouse. This took about three days. To help you get into the spirit of the thing, this was in January, a particularly cold and wet January. On the morning of the third day, a dear backhoe driver named Bob just happened to be passing up the driveway to do some septic work. I must have been a pitiful sight because he stopped and motioned me out of the way. In a mere five minutes, he had completed the task, squared off all my edges, and even removed the middle so that we had a flat, empty square where we could begin building the form.

I worked with a carpenter for the next part of the project. The form was built with two-by-eights, three high, which we braced with stakes every two feet or so. I remember thinking that all

Early spring seedlings are spurred on by the incredible warmth of the glasshouse. As the sun's rays get stronger, seedlings are moved outside or to a cold frame.

that bracing was extreme overkill, but when the cement truck came and the form started filling with cement, we had some tense moments thinking the boards might blow out.

Once the cement dried, we took off the forms, put some perforated PVC pipe around the foundation to help with drainage, and then filled it all in with drain rock. Unfortunately, I hadn't communicated properly that I wanted a soil-floored greenhouse and not drain rock, so it was more expensive than it needed to be: I ended up just covering the rock with soil so that I could grow plants on the floor.

Then it came time to assemble the thousand or so pieces of the actual glasshouse. It took me about ten minutes to decide that there was no possible way I could figure it out. Fortunately, the glasshouse had been made by a local manufacturer who was still in business. BC Greenhouse came to the rescue and the whole thing was assembled in a morning.

After we finished building, I realized a fatal flaw. I had set it up only a stone's throw from the road, along which dozens of children walk to the nearby school. I figured I had erected an unbearable temptation that would only lure young boys and girls to take aim. Thank goodness, not one child, or adult for that matter, has ever thrown a stone. This might be because they all saw me toiling away, digging the foundation . . .

The Recycled Wood Greenhouse

A commercial nursery down the way was closing, and they were selling off their twenty-year-old wood-frame greenhouses for next to nothing. The deal was that you had to take down and haul away the hundred- by thirty-foot structures.

Always game for a bargain, my then partner and I jumped in wholeheartedly. Unfortunately, I hadn't realized that the base plates, on which all the trusses sat, were treated wood. This is not allowed by our organic regulations, so we had to replace them, and opted for cement footings.

Since we were dealing with quite a slope, and we had decided on footings, we chose not to level the site but to level the footings. This was a very time-consuming undertaking, and if I could do it again, I would choose metal footings (like Long John, see page 132). We decided that we wanted to raise the greenhouse a bit higher so that we could capture more hot air in the structure. However, the geometry of the situation got ahead of us, and when the slope was factored in, we ended up with a thirteen-foot-high greenhouse, instead of a nine-foot one.

Building a structure this tall proved to be more of a challenge than I could handle. Again, to get into the spirit of it, it was another January; we had started in October but that is how long it had taken. We required scaffolding to join the half-trusses in the middle. We would abut four half-trusses onto the top plate, set their heels into the cemented footings, and then hoist it up in stages until I was on top of the scaffolding, holding the set above my head at the right level. Richard then nailed the footings in, and connected the trusses to the rest of the spine that we had already erected. Images that stay with me from the process are heavy wet wood, perching precariously on two-storey planks, and being very cranky because progress was always being hindered by wood that needed de-nailing, rotten bits that needed cutting out and replacing, and warped bits that needed reorienting, et cetera, et cetera.

Salad greens in succession in the greenhouse. After two cuts, the bed is scythed, fertilized, rototilled, and replanted immediately.

On the plus side, it is definitely the coolest-looking greenhouse around, and I am very proud of it. It is a great community conversation piece, and has served us really well. We covered it with a heavy grade of vapour barrier that lasted three years. It was a fine substitute for very expensive greenhouse plastic while we were honing our greenhouse-growing technique. When you get your greenhouse up to optimum performance, and every little extra bit of light you get equals big bucks, then it is well worth investing in the clearer plastic.

If you really enjoy the process of building, and the creativity it requires to use recycled materials, then by all means, it is a worthwhile project. However, if you just want somewhere to grow food, save up and buy a metal greenhouse kit.

Many eyes making sure that the ridge line of the new greenhouse is level. This ensures the structure is square and that the sheet of plastic will cinch up nicely without bunching.

Long John—The Aluminum Kit

I wish I could relate a story of a single after-noon's work party resulting in the erection of a lovely prefab structure but this is not our story. A few years ago, Rachel, Heather, and I bought a hundred-and-sixty- by twenty-foot aluminum greenhouse from a company called Harnois, in Quebec. We were assured that this was the Cadillac of all greenhouses. Unfortunately, Harnois was better at engineering greenhouses than at writing directions, and the diagrams mystified us.

We had gone together with several other farmers to purchase several greenhouses at a bulk rate. The pieces all came, more or less clearly labelled, on a big transport truck. The first step, although it is truly the last thing you want to do, is to take a careful inventory. Most companies only give you a certain time in which to report damaged or missing parts. One of the farmers in our group was missing quite a few parts, which he failed to notice within the time window, and he had to replace them out of his own pocket.

Next came the levelling of the site, which was done by tractor. Even a slight slope becomes quite significant when you are levelling, and we ended up with a water-catchment area on the uphill side of the greenhouse. This has led to perennial flooding problems during winter rains and has required a drainage ditch outside the greenhouse. If you need to level, think carefully about the topsoil. Levelling the site helps square the foundation but in order to level out a slope, you inevitably have to scrape off some topsoil. If you decide your site needs levelling, scrape off the topsoil, put it in a separate pile, and then do the levelling. Replace the topsoil evenly afterward.

Once we had the site prepared, we began to hammer in the posts. This time we were working in June, just after the soil had dried up. We had a problem in that the soil was very rocky and we kept hitting rocks that would knock the footings off-centre. Some posts wouldn't even go in all the way which was frustrating, because having everything level is important. We ended up correcting our error by adjusting all the posts to match the one that couldn't go down any farther. By doing this we compromised the structure's anchor but we just had to get on with it.

We were so busy with our own farms that we had little time to devote to this project; we resorted to Sunday evenings, 6:00 PM until dark. Each week we would forget what we had figured out the previous week, so we ended up wasting a lot of time. We finally got the last door on in mid-November.

Having a new, well-engineered greenhouse has really made clear what the old wood greenhouse is lacking. For one thing, the shape of the trusses is such that it allows lots of headroom right out to the sides of the greenhouse, so there's room to manoeuvre a rototiller through the whole area. The pitch of the roof is better for managing snow load, but we still have to keep a close eye on it.

Another great feature is the roll-up sides with gears. The sides roll about a third of the way up the side walls of the greenhouse to allow air to circulate. Greenhouses can get too hot in the summer, which can cause all sorts of trouble. Roll-up sides are essential, and having a safe gearing system is a nice touch. Old-school models of roll-up gears were under so much tension that if you let go of the crank, they could smash your knuckles. After all the glitches, we are very happy with our Harnois greenhouse.

Pros and Cons of Greenhouse Growing

Greenhouses are a boon to any farm. The climate conditions that you can achieve in a greenhouse, namely spring warmth and summer humidity, can provide perfect growing conditions. Greens grown under plastic have a luscious, velvety finish that restaurants love. Greenhouses also keep plants dry in the late summer rains that can cause blight and other catastrophic diseases. Depending on your soil and climate conditions, and your ability to vent the greenhouse, you may run into challenges, such as mineral build-up, with greenhouse production.

Mineral Buildup

One spring, maybe four or five years after we built the greenhouse at my farm, I noticed a whitish crust on the surface of the soil. It reminded me

Rachel's Water Breaking

We were almost done with L.J., the greenhouse. After several months of work parties, we were meeting to put on the end doors so that the salad would be protected from the winter cold. It was just Robin and our friend Greg who showed up to work: Heather was away and Rachel was nine months pregnant. Greg mentioned to Robin nonchalantly that he had heard Rachel's water had broken. Rachel had been having problems with her well that year, so Robin sped up with the work on the door, shaking her head and thinking that she would hurry over later to help. "Just what she needs: nine months pregnant and now her water main has broken." Did she ever feel silly when she found out that Rachel had left for the hospital because she was in labour.

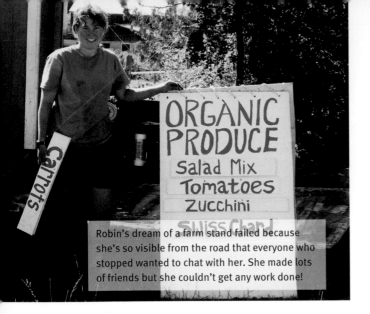

Robin's dream of a farm stand failed because she's so visible from the road that everyone who stopped wanted to chat with her. She made lots of friends but she couldn't get any work done!

Drainage Ditches

My farm is in a low-lying area. I had put up with water pooling on the land in the winter for a long time, until Andrew persuaded me to try installing drainage. I was hesitant to shell out money for something I couldn't imagine working. We started the first year with a hundred-and-twenty-foot pipe at the very top of our field which we connected to an existing length of drain tile (a segmented clay pipe that is laid in sections and used to drain water out of wet areas). We had a backhoe come in and dig a sloping trench two feet wide and three feet deep. We put a four-inch layer of drain rock on the bottom and then a four-inch perforated PVC pipe (perforations facing down onto the drain rock). We had to test the grade to make sure the pipe was properly sloped. We then filled the trench to the top with drain rock. We saved money by narrowing the trench on top of the pipe; we divided it in half vertically, filling half with soil and half with drain rock. The point is just to bring the drain rock to the surface so the water can funnel down through it. The trench paid for itself in one winter with the increased production in the greenhouse, because the soil remained dry.

The next year I did the entire periphery of the field in more or less the same way. One fatal error that I made was underestimating the capacity of the existing drain tile. We connected two pipes to the drain tile in a T-junction, but it couldn't handle the increased volume during heavy rain. The water overflowed the trenches and ran across the land, causing massive erosion. The next year I had a wider pipe installed next to the drain tile so now there's more than double the capacity.

a lot of the salination that I had seen in tropical soils. I had some standing water in the greenhouse over the winter, because I have a high water table, so I figured it was just residue. It "went away" when I watered so I tried to ignore it but when it was worse the following year, it was hard to keep pretending it wasn't an issue. Concurrent with this problem, I noticed a drop in tomato production. I thought the plants were dropping their flowers because of the extreme heat so I bought hundred-foot roll-up sides with a geared crank. These were a good purchase because they allowed us to vent properly, but they didn't solve the blossom drop.

I installed drainage (see next section) and that stopped the bulk of the buildup. What was happening was that the water was evaporating off the top of the soil, leaving the salts behind, and we were never watering heavily enough to rinse them back down. We finally decided to cut the plastic off and let the soil be cleansed by the heavy winter rains. This worked well, but the final solution was to install drainage ditches so I could prevent the pooling altogether.

I think the soil is a lot healthier, thanks to the drainage. I also feel it paid for itself in a single season because it increased my capacity for extending the season.

Harvesting

A well-organized packing area is a very important part of the farm's function. You should have tables to work on, and a large tub or two for cooling and rinsing produce. I have two old bathtubs that work well. The packing area should be covered, so that you can keep the produce out of the sun. Make this area bigger than you think you need, to allow for increased production in the future.

Sharp knives and scissors are essential for harvesting. If you find yourself misplacing them in the field, try tying bright flagging tape to the handles. Or, like me, you might just choose to have so many scissors and knives that you stumble across them wherever you walk. I use Rubbermaid bins for most harvesting, except tomatoes which are harvested into cardboard flats. We have standard sizes of bins, with the weights of each recorded near the weigh scale.

There is a real art to harvesting properly, to ensure that the produce looks nice and stays fresh, and we have an order of operations for our harvesting routine. We're always pressed for time, and we want the produce to arrive as fresh as possible. Our orders for restaurants and the box program have to be filled by Monday afternoon, so we start on Sunday. We start with root vegetables: beets, leeks, onions, turnips, and carrots. We generally keep the tops on everything, so handling is important. We have a trigger nozzle on the end of a hose that we use to blast the roots clean while they are on a table. We lay them in a single layer, and flip them once or twice before plunging them into the bathtub for a final rinse and to clean the tops. Discoloured leaves are discarded and leeks are trimmed with three well-aimed snips.

We start Monday morning with lettuces, salad greens, and leafy vegetables. These are the most fragile, and benefit most from being picked in the cooler hours of the morning. Cut them, and let them sit in cold water for about ten minutes, but not longer. You don't want them to absorb too much water, which will burst their cells, but you do want them to lose all their residual field heat in the cold water. It is paramount that greens be kept cool, moist, and out of the sun and wind.

Submerging greens in cold water helps draw out the "field heat." Having refrigeration for them further increases their shelf life and appearance.

Harvesting salad greens is done with scissors. The greens are planted in blocks because they grow at slightly different rates.

Tips for Growing Salad Greens— Robin

Salad greens are a good money-maker on our farm, and an important way in the door at many restaurants. Once you're delivering salad, it's easy to persuade a chef to add other produce items to the order. One challenge with greens is having a consistent supply. Our salad greens system is based on a tight succession plan so we don't have to cut a main season salad bed more than twice at the most. Most salad greens can be grown as a "cut and come again" crop. However, repeat cutting takes lots of time, because the regrowth isn't as thick and even, there are damaged leaves from the first cut and the bed has more weeds. Flavour can also suffer, as the mustards tend to get spicy, and lettuces and endive can get bitter.

Each spring, I figure out roughly how much salad I can sell in a given week. In my case, this includes three harvest days per week. I then estimate how much area will yield this amount. For example, on my farm, each hundred- by three-foot bed should yield fifty pounds of perfect-sized leaves. I aim to sell fifty pounds a week, so my salad section includes four beds of this size. I plant the beds in one-week intervals, and till each one in after a week of harvesting. This means I always have one bed just-seeded, one growing, one primo bed for harvesting, and one from which to scrounge a bit of second-cut salad in case of larger orders. The key is to be disciplined about the timing. It is often tempting to leave a bed a bit longer, and scrounge more second- or third-cut salad, but this throws off the rotation. The greens we include in our salad mix are mustards (mizuna, ruby streaks, giant red, southern giant), chards (bull's blood and rainbow), red Russian kale, arugula, endive and lettuces. I find all the greens except the lettuces do quite well in a winter greenhouse until the weather plunges

Andrew steps over a perfect bed of kale. During his time at Feisty Field he revolutionized the way our salad is grown.

to below -10°C. I do a consistent rotation until mid-September, at which time I try to have all the beds planted so the greens will have a strong start before the weather changes. Mustards don't get as spicy in the cold temperatures, so I'll harvest them six or eight times before I start reseeding in mid-February.

Greens grow best in finely tilled soil with high and even fertility. It is essential to maintain high fertility so you can have a predictable and even supply of salad. I work in small amounts of fish compost, about two or three wheelbarrow loads in a hundred- by three-foot bed, with each seeding. I try to maintain a pH of about 6.8 and consistent levels of phosphorus and potassium, which you can gauge with a yearly soil test. I find they need regular, light watering. Because they are seeded so thickly, there is a danger of them rotting if they are too wet. To seed a bed, I make four shallow troughs running lengthwise down the bed. I sprinkle the seed between my fingers like spice, dropping it into the furrows from a standing position. I barely cover the seed using the back of a rake. The first watering is heavy, but once the seeds have germinated we maintain light moisture. Greens are always sealed under floating row cover to prevent flea beetle damage.

A story that has gone down in our collective history is about an epic order for a hundred heads of pac choi. I had never sold such a quantity and I didn't have enough bins. I was so proud of the dark green heads that I didn't want to crunch them into bins anyway, so I left them open in the back of my pickup and drove to town. When I arrived at the warehouse, the poor pac choi were wilted and so sorry-looking that I was ashamed to deliver them. Just thirty minutes earlier, they had been glowing with vitality. This was a good lesson in why you should always keep greens sealed in a bin! The happy end to this story was that the kind buyer offered to revive them for me by plunging them in ice water, so they were saved, but I had a good cry about the whole incident later that evening with Heather and Rachel.

Lettuces can be submerged and inverted as you grip them by the base. They can then be swished around under water with moderate vigour to get the dirt out from inside. Salad greens in volume are best washed with a two-tub system. First they are spread on the surface of a half-full tub in a single layer where they are rinsed and picked through. They are then transferred to a second tub in small quantities for a final inspection before being put in a bin. The bin, with its lid on, is then set on an angle to drain and the water is strained out by tipping the bin periodically.

Beans, peas, and cucumbers are picked next. It's best if you don't wash them. We put these into plastic bins immediately and set them in the shade. Long English cucumbers are especially prone to wilting, so seal them in plastic bags and refrigerate them, if possible.

Berries are last because they are fragile and can't be picked wet with dew. Do not wash them. Our customers understand that we can't guarantee delivery of berries, because they won't meet our standards if the weather is either rainy or too cold. They don't sweeten properly without the heat, and they don't store with even a bit of dampness. When that's the case we pick them anyway, cut off the stems, and freeze them immediately. We were pleasantly surprised to find that berries that didn't meet our sweetness standard in the summer were delicious in the winter. Everything's relative.

Be attentive to the needs of different varieties: pale-skinned cukes bruise each other; don't stack heirloom tomatoes; don't move summer squash around too much because they scratch. Make sure you're always tasting the produce. This will prevent you from arriving at market with woody beans, bitter lettuces or overly pungent mustards.

First Market Day

I was up at the crack of dawn to harvest my first offerings for the Moss Street Community Market: kale, chard, arugula, radishes, baby spinach, and salad greens. I skipped through the rows collecting all my beautiful vegetables; I scrubbed, bunched, polished, and bagged. I was thrilled to be going to the market. When I got to the school grounds where our market is held, I set up my table, placed everything lovingly in baskets, and wrote out the prices on my little chalkboard. Then I took a breather and walked around the market before it started.

My bright, shining smile started to droop as I looked at everyone else's stands. They were

A gorgeous bed of late-season Scarlet Runner beans.

beautiful! "How did they grow all that variety? And everything looks twice as vibrant as mine." I returned to my puny table and made a resolution: I was going to stand behind that table and not cry. I was certain that nothing was going to sell and I was a dreamer for thinking otherwise. I was out of my league. All these farmers had been coming for years, and they were experts. Customers knew them and were going to buy from them, of course. When the time came to pack up, I would just load everything back in the truck and never, ever come back.

When the bell rang to announce the beginning of the market I was stunned as people sauntered by, stopping to "ooh" and "ahh" over my baby spinach. Everything was gone from the table in less than two hours. During the rest of the market, I wandered around and bought all sorts of delightful treats. I got to talk to the farmers, and related my experience. I got some very supportive hugs and encouragement. Over the years I've come to really value my relationships with the other market vendors.

Chickens for Meat

Our friend Lana moved to a rural property and took to farming with her whole heart. She built the sweetest little chicken pen and she was very proud of her small flock of layer hens. When she was out in the garden one day, to her horror she saw a bald eagle swoop down and take a chicken. She must have been a threatening sight, because the eagle dropped the mangled bird before flying away. Seeing its innards turned outwards, she realized that she was going to have to put the chicken down. After some nervous deliberating at the chopping block, she decided to take it to her neighbour to do the deed. He was an eccentric old guy who took one look at the chicken and declared that all she needed was some surgery. Lana watched in dismay as he shoved the innards back in and sewed up the chicken's front with rough stitches, feathers and all. This seemed like torture to the bird and a real long shot for recovery.

She reluctantly brought her dear little bird home and gently put her in the henhouse for

Elias does a close inspection of the three-week-old chicks just put out to pasture.

the night. The next morning, when she was still in the same spot, Lana carried her to the water dish. She drank eagerly, but the water dribbled out from the sewn-up area. Thoroughly traumatized, Lana took the hen back to her neighbour and demanded that he put her down this time. Instead, the guy got out the thread again and put in a few more stitches. The next day, the chicken was as good as new and lived another three years or so. Lana named her Lucky!

I have also had a few forays into livestock and farm animals; none of them were very profitable, but they are definitely a source of great stories for friends and family. In the early years, Rachel and I decided to raise broiler chickens. Being vegetarians at the time, we knew less than nothing about raising chicken. I had never even had a pet before.

However, we had read about sustainable-farming systems, and they all recommended incorporating animals into rotations and creating on-farm fertility with animal manure. We decided to try pastured poultry, the chicken tractor model, which involved a movable pen right in the field, where the chickens could scratch up beds and prepare them for planting. We called another farmer (who took us under her wing) for advice and she assured us that raising chicken was a profitable endeavour. Although she had never tried the movable-pen system she gave great advice and ongoing support.

We typed "chicken tractor" into a search engine and downloaded a picture of more or less what we ended up building. We had many laughs putting that structure together; it still wobbled, even with gussets and cross-bracing on every corner and joint. Rachel ordered the chickens through the mail. She set up a cardboard enclosure in her basement that was about six feet around with some wood shavings for bedding. She hung a heat lamp in the middle, added a standard chicken water dispenser, and used egg cartons to hold the feed.

Sometimes your decisions are just inconsistent with your values but you don't realize it at the time. We decided that we would raise Cornish Giant hens, which are the standard in the meatbird industry. These over-bred monsters are eating machines. They convert grain to meat at astonishing rate: after six weeks they weigh five to six pounds and are market-ready At $3.50 a pound, we figured we would be making great money with twenty-five birds. We were floored when the four-inch-tall chicks powered through a fifty-pound bag of feed in a little over a week. They were really cute at first. I remember leaning over the pen and getting a game of duct-tape football started by tossing in a fragment of tape, and watching them all run for it, and pass it around. By the second week, they had lost their baby fluff and started to grow their white feathers.

When it was time to move the chicks out to the field, they really didn't know what was happening. I had heard that chickens were supposed to till the soil like little tractors, but all these did was compact the soil by sitting on it. What a joke! I tried to train them to dig, by turning the soil in front of them. They would shuffle around a bit and grab worms that were within reach, but basically they just wanted to chow down on the grain. They grew daily, and their huge thighs and lower legs reminded me of dinosaurs as they lumbered around when I scooted the pen forward. I have since learned that the chickens you want

for tilling are layer hens, which are more agile and have higher energy levels, which enables them to scratch for food. While I was caring for the growing chickens, Rachel had ordered another batch of chicks that would be ready for the field when the first batch was slaughtered.

The visit to the slaughterhouse was pretty traumatic for us newbies. I don't know what we were expecting but there was nothing muted about the experience. I'll spare you the details but after the job was done, we looked at each other and exchanged a silent, "Oh no! We still have the next batch of chicks to deal with." We got lots of compliments on the chicken, and we even ate some ourselves, but after we considered all the hours involved and the costs, we decided we wouldn't do it again.

Raising chickens for sale on our scale is not profitable because of the cost of both certified organic feed and slaughtering. We are currently facing a regulatory regime that is impractical and prohibitively expensive.

Rachel checking on the feed and water in the larger fenced area for her meat birds, now eight weeks old. She locks them up in the pen overnight for protection from predators.

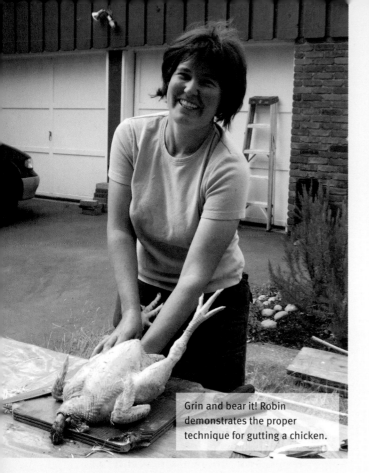
Grin and bear it! Robin demonstrates the proper technique for gutting a chicken.

Meat Regulations

A food-safety regime is being forced upon small farmers in Canada, partly as a response to consumer fears about contamination. The irony is that it threatens the very sector that is possibly the safest: the small farm that markets directly to consumers. The Food Safety Act that was passed in 2004 specifies meat-inspection regulations that, among many requirements, necessitate pre- and post-mortem inspections, as well as slaughter in an inspected facility. These standards are appropriate for factory-sized slaughtering facilities where workers don't know the animals. Even then, with all the regulations being followed, Canada still ended up with a listeriosis disaster. In 2008 and 2009, this deadly bacteria was found in a meat-processing plant in Ontario. Since this single plant services the whole

country, we saw how centralized facilities create fragile food systems. Although the meat processor had been following all the current regulations, these lethal bacteria kept surfacing in meat all over Canada. The plant had been running continuously for decades, and the conditions within had been inadvertently selecting for resistance in bacteria. A daily regimen of applying approved bleach-based sanitizers and the accompanying cleaning procedures had actually accelerated the growth of resistant bacteria. The factory mentality, taken to extremes, overrides nature's inherent ability to cope with pathogens.

Small-scale farms are not immune to food-safety problems, but they have a few built-in checks. Farmers who raise and slaughter their own animals are aware of health defects and can easily sort those animals out of the system. They know when the animals have had their last meal, and can slaughter them with empty guts (a full gut is a major source of contamination). They can process in small batches, and they have control over every stage in the procedure. Finally, many small farmers have long-term relationships with their customers. Every meat farmer I know has a regular clientele and a waiting list. The farmer knows the families she is feeding, and this relationship can last many years. There is a direct feedback mechanism and possibly some caring that goes into every economic transaction. This system reflects a circular pattern in which customers return to the same farmer, unlike the linear pattern of a grocery store commodity purchase in which the particular farmer is out of the picture. If the food from a small farm happens to be contaminated, a recall can happen quickly and thoroughly. These farmer-customer ties are based on a deep sense of trust and goodwill that cannot be replicated by inspection. The farmer's livelihood and reputation is riding on every transaction.

However, Rachel now raises sixty birds a year for our personal consumption. It is worth the effort for us, because we value chemical-free and ethically raised meat. Interestingly, slaughtering the chickens at her farm, with friends, is not traumatic like our first experience was. We felt that taking responsibility for this aspect of the process was the right thing to do; we slaughter the birds with the utmost respect and care for their well-being. It feels important to have this skill.

Lambs

During my early years of farming, I worked part time as a field hand on a mixed organic farm. One day I was invited in for a meal, and they were serving their own lamb. I was a vegetarian because I disagreed with the industrial-meat industry, but I couldn't find fault with the meat that was served to me, so I just tucked in. It was so delicious that it was a transformative experience. I reckoned that I couldn't ever afford meat like that, so the only way I was going to get it was to raise it.

I went to the library to read up on sheep because I wanted to know how to choose a lamb. The book explained how to look for size and shape, and most importantly, to never, ever buy a bottle-fed lamb. Bottle-fed lambs didn't have the immunity passed through mother's milk and, since they had not been accepted by the mother, there was probably something wrong with them. Moreover, they were just too used to humans, which would lead to problems. Armed with my library book, I felt ready to go into shepherding.

Richard, my partner at the time, was from New Zealand, and he knew a few things about

Rebecca showing the next generation how to clean chickens.

sheep, or so I thought. When we got to the sheep farm, the farmer put on a show with her sheep dogs, and soon the new lambs were lined up in front of us. I started walking down the line, looking at the build, stance, and shape of the lambs. When I glanced back, Richard was ringed by bottle-fed lambs: two ewes and a runty ram, suckling his fingers and snuggling his calves. His sheepish smile was a classic. We ended up driving home with three lambs on his lap in the cab of the truck, rather than in the back canopy as we had planned. What a mistake!

Before getting the lambs, we had divided the fifteen-foot-wide buffer zone between the peripheral fence and our cropped area into hundred-foot runs with recycled, four-foot fencing staked into the ground with rebar. The first lesson was that our spring clover and timothy grass mix was too rich for the lambs. Puku, the runty ram, was particularly susceptible to the gas created by the nitrogen-rich grass. His belly literally blew up like a beach ball, and then he'd fall over. He would eventually deflate, but when I got him to his feet

Robin and Puku, the free-range, strawberry-fed lamb. Raising lamb was a mostly wonderful experience but very impractical without proper fencing.

again he would immediately start munching, so I'd have to lay him down again. I'd visit him periodically, because he couldn't pee lying down. I'd right him and let him pee, but then I'd have to wrestle him down again. Once he deflated, he'd hop up again and happily go on eating, never learning from his experience.

I built the lambs a little shelter with a corrugated tin roof. One morning after a windstorm, I got down to the farm and Puku was missing. The tin was flat on the ground. I searched high and low for him, and finally, when I called him, I heard a muffled "Baahaaa." He had been caught under the very thin piece of tin in a small hollow, and he was quite relieved when I lifted it up and freed him. To grasp the full humour of the situation, you have to appreciate how tiny the piece of metal was. What a runt! But I loved him.

Richard had assured me that lambs didn't

drink. "Can you imagine farmers in New Zealand wandering their thousand acres with a water bucket?" he asked. I tried taking the lambs water that spring, but he was right: they wouldn't drink. Then, while visiting the lamb section at a fall fair, I saw, to my absolute horror, lambs drinking from buckets in their pens. I literally ran from the fair and drove straight home. I took a bucket out to my lambs, but they looked at me with that blank stare that only lambs can perfect and kept on grazing. As I sat on the grass, perplexed, I realized that the lambs I had seen were probably grain-fed lambs. In our grazing rotation, we gave the grass a good watering after the lambs had finished on it, and as a result, the grass was lush and green, much like the grass in sheep-raising areas.

One thing we learned shortly after getting the lambs is that they are escape artists. No matter how far I jammed the rebar into the ground, or how many fencing staples I used to secure the bottom of the fence, they always escaped. They broke the fence to get out, and they'd break it again to get back in. It was probably torture for them to be fenced next to veggies and strawberries, but I've heard from other people that sheep need good fencing because they are incredibly strong, and they can slither through any gaps.

A very interesting social feature of lambs is that they need to be around a bellwether. The bellwether is a natural leader, or in Puku's case, the animal who has a certain magnetism that attracts the others. Whatever the reason, Puku was the bellwether, and everywhere that Puku went, the others were sure to go . . . If I was moving them, I just had to focus on getting Puku into the pen, and the girls would follow. If the girls broke out, and Puku didn't make it through the fence because of his rotund shape or accident-prone manner, the girls would break back into the pen, by another route if necessary, in order to be with him. Once I figured this out, managing them was much easier.

On the day the lambs were slaughtered, Paul, my apprentice, found me teary and glum. He was surprised, because the whole season he had thought I had a grip on the fact that I was raising the animals for meat. I really did have an animal-husbandry mindset, which, for the uninitiated, seems paradoxical: I loved those animals even more because I knew I was going to kill them in the end. I really cared about all their life stages. I wanted them to be comfortable, secure, and well looked after at all times. (Okay, I did want to string them up by their hoofs when I caught them eating my strawberries, but that was just a fleeting thought.) I spent time researching the most humane slaughterhouse because I didn't want them to suffer. Since I was responsible for their existence, I wanted it to be as good as it could be. I was sad when they were gone, but I ate them and truly enjoyed the meat. I don't think I'll cry when I eventually raise my next batch of lambs, but I think that my respect for the animals and my appreciation for their meat will never diminish.

Vegetarianism

As Heather, Rachel, and I were discussing our commonalities and differences, an unexpected one surfaced: we are all reformed vegetarians! As beginning farmers, we were all young, firmly committed environmentalists, and at that time, our environmental commitment included vegetarianism. As our farming careers have progressed, and as, we feel, our environmental commitments have deepened and become more nuanced, we have incorporated meat thoughtfully into our diets. This has come from an understanding of how animals can be an integral part of the farm system. Animals eat pests and unsaleable crops, and their manure feeds the soil. Sheep can be kept in marginal areas that are unsuitable for vegetable production, yet eat grass that would otherwise have to be mowed. Chickens can be moved through a vegetable rotation, providing fertility and breaking up disease and pest cycles. At the same time, animals can be raised humanely, live a life that is longer and safer than possible in nature (as if domestic farm animals ever would have evolved without human intervention!). Yes, ultimately the animals are killed, but this can be done respectfully, quickly, and with minimum stress and pain.

Certified Organic: Is it for You?

I didn't really think over the pros and cons of certifying before I applied, because I had been seriously indoctrinated into the philosophy behind the movement during my apprenticeship. I have never regretted certifying, although I respect those who opt out of the certification process. The movement for "beyond organic" has gained its fuel from the tragic watering-down of the organic standards in the United States in 1998. As British Columbia farmers, we have been fortunate so far to have ownership over our system, although this too could change. We are subject to a soon-to-be released national standard, but we have some freedom within the structure and we hope that it will evolve differently from the American process.

We are lucky in our area to have a local certification body comprised of farmers who really care about the integrity of our agriculture. They have put together a set of guidelines that are locally relevant and thoughtfully presented. I knew the organic guidelines before I started the farm so I didn't have to change to suit them, and I found the organic farmers to be a welcoming community with whom I shared many values. They were open to sharing their techniques, sources for materials, and favourite plant varieties. They had good suggestions for marketing, and had even developed a listserv to help chefs and farmers connect. I feel my yearly fee of about $400 is worth it, just to be part of this network of growers.

The certification inspections are thorough but they are meant to be a teaching experience rather than a punitive grilling. Some of the guidelines can be open to interpretation, and the inspector makes sure that you are truly on the same page as everyone else. Some regulations may need to be broadened, and all the organic growers are encouraged to be part of the guideline-revision process. The inspection is also your opportunity to explain how and why certain guidelines aren't working for you, and suggest alternatives.

Conclusion

There are many more stories I haven't told about this adventure into farming, and there will be many more, I'm sure. As I start my eleventh season on this land, I'm excited about having a business partner—my farmhand from last year, Dennis, whom I'm inviting to be a part of running the farm. This is an exciting step—opening up and managing the farm for the benefit of two separate households. I think this is a strong trend in farming: away from the nuclear-family model and toward partnerships and various other arrangements that help agriculture fit with the realities of our modern lives.

I have great hope that new forms of agriculture will breathe new life into the old model and create space for different kinds of relationships to land and to food. I am inspired by community models like land trusts and the Community Farms Program that are allowing these new forms of farming to unfold in British Columbia. I think it is an exciting time to be in farming. Food sovereignty theory is infiltrating government policy, which is wonderful news.

Change is coming and farmers are leading the way.

Yugoslavian hardneck garlic is Robin's favourite variety because it has strong, pungent flavour and large, easy-to-peel cloves, and it grows really well on her farm.

On Being a Woman Farmer

It was eye-opening for me to learn that farming in most parts of the world is considered women's work. When I'm working in the fields, it feels like the most natural thing in the world for a woman to be doing. The nurturing of life and the aesthetic of the surroundings of the farm feels very feminine. Harvesting, and the way a bin fits so nicely on a female hip—that seems natural. Seeing women on tractors is now common, but when I watched my first boss, Fran, on her tractor, it was very empowering and liberating to know that anyone can learn these skills.

At first I felt limited by the idea that I was a woman in a physical occupation. I felt insecure walking into a building supply store, or talking with a machinery operator to describe what I wanted done. I felt that if I couldn't figure it out myself, it was proof that I shouldn't be farming. My confidence has grown with age and experience, and now I know that there is a tool for everything and that, of course, even men don't always know how to do a job (no matter how confident they sound). I feel as though I don't have to prove myself anymore because I've succeeded, and this confidence has flowed into my personal life. I don't

hesitate to do my own plumbing and refrigerator repairs, even some truck repairs. It's like I've broken down a psychological barrier between what women should and shouldn't do, and now I can try it all. I find men very helpful and respectful when I seek more knowledge, and this keeps me inspired to learn more, do more, and bring more women along with me.

I love mentoring women into farming because I love to empower them in non-traditional roles. I teach them how to be safe using power tools, I give them opportunities to draw up building plans and to shop for construction materials. I still chuckle when I think of the look on one apprentice's face when I said, "Well, you don't want to have to wait for a man when you need your greenhouse up and ready." She looked both shocked and excited; it had simply never occurred to her that this was something she could do on her own. I love that the children on the farms are growing up watching women in unconventional roles. I think that if they are accustomed to seeing women and men working together as equals, they will be less bound by gender roles, and the potential of that is exciting.

CHAPTER FOUR

Why Organic?

Shelling beans is one of many farm tasks where kids love to participate. They are mimicking a task done all over the world, as dry beans are a staple food for much of the world's population.

Growing food is just a part of it. We can't separate our motivation to sow seeds, water, weed, spread compost, harvest, and eat the result of all this from certain ecological, social, and political issues. These undercurrents inform our beliefs and give our choice of livelihood a deep meaning beyond simply the satisfaction of growing food that nourishes our bodies, our families, and our communities. A recurring theme in this book is the hope that what we are doing will have some impact beyond our region, that we are contributing to a movement that is echoed in many forms across the continent and beyond. What follows is a brief exploration of some of these political, environmental, and social issues. These are topics about which many books have been written, by people far more expert than we are but we hope to pique your interest and inspire you to read more (see our Bibliography).

Seed

To us, the term "food security" evokes a sense of threat as well as a sense of hope. Food security can be defined as "the availability of food and one's access to it." Increasingly, we are seeing food systems being dominated in powerful ways by corporations. For example, the concept of patenting a seed would have been completely abhorrent to farmers in the not-so-distant past.

Seeds and humans have co-adapted since agriculture began, some ten thousand years ago. The favourable response of seeds to human intervention is a testament to the wonderful complexities of natural adaptation that enable them to survive and thrive. We select seeds for the traits we value, such as size, taste, digestibility, and tolerance of different soil types and moisture levels. Seed has been traded, shared, and moved around the globe to be tested and enjoyed in different environments. It has responded well to human manipulation. Our existence and survival are bound up with the existence and survival of seeds. Think about what it means when corporations can patent the seeds we depend on for survival, and penalize those who grow the seeds without paying the licensing fees. When Monsanto Corporation bought Seminis Seeds in 2005, it was estimated that they took control of 40 per cent of the vegetable seed market in the United States and 20 per cent worldwide.[1]

Big seed companies develop seed varieties that are suited to industrial-scale growing, harvesting, and transport. They put little or no effort into preserving heirloom varieties or developing new varieties that would be well adapted to small-scale growing in different bioregions. They select for appearance, uniformity, ease of harvest, and storage, not for taste, resilience, and nutrition. For thousands of years, farmers around the world have contributed to developing a rich heritage of seeds. The corporatization of seed is threatening this heritage and resulting in the incredibly rapid loss of crop diversity.

Rachel's son, Elias, at an agricultural rally. She suggested that he write about what farming means to him.

In this increasingly industrialized world, we small-scale organic farmers can feel as though we are moving against the grain. A twenty-two-year study headed by David Pimentel of Cornell University compared the economic, energetic, and environmental impacts of organic and conventional systems involving corn and soybeans.[2] One of his conclusions was that the yields were close to the same in both systems but that the economic return was higher in the organic system.

The organic system required 35 per cent more labour, which was spread out over the whole season; the farmers participating in the study could perform a fair amount of it. In the conventional system, the labour requirements were heavy in the spring and fall, which necessitated more hiring of labour by the farmers. Thus, the hired labour costs in both systems were roughly similar. Four herbicides were applied in the conventional system, two of which were subsequently detected in water samples. One of these, atrazine, exceeded the Environmental Protection Agency's maximum contaminant level. The organic system used 30 per cent less fossil fuel and, through the use of cover crops and organic inputs, actually increased the soil biomass and soil biodiversity, and contributed to water conservation. The study concluded that organic faming methods reduce the negative impacts of agriculture on the environment, and do so with yields comparable to conventional practices and a good financial return.

The general public seems to believe that large, highly mechanized farms are more efficient, and have higher productivity than small farms. An increasing number of studies, however, are concluding that small farms score the same, or higher, on both counts.[3] Studies into the interactions around small farms are also concluding that they are agents of "food justice."[4] Food justice is a wide-ranging idea. It encompasses both the rights of consumers to nutritious, accessible food, and the rights of farmers to have control over their agriculture. Food justice also considers the ecology of food-producing regions. Food justice is a major component of the larger vision of food sovereignty. With industrial food systems, there is little opportunity for food sovereignty.

Patenting a Seed?

Patent a seed? A seed is not an artificial product. It is the height of arrogance to assert that by changing a single gene in a seed, that seed can become private property, when for countless generations humans have been breeding seeds for the collective good. What kind of food security do we have if a seed, created in nature, can be tweaked in a lab and suddenly become owned by a corporation, causing the farmers who have used it for generations to have to pay for its use? And indeed, to pay penalties if it turns up on their land through ordinary means such as cross-pollination? Percy Schmeiser is a Saskatchewan canola farmer who was sued by Monsanto Corporation in 1998 when the company discovered some of their genetically modified Roundup Ready canola growing on his land. Schmeiser claimed that he had not grown it intentionally, but that the seeds had blown in on the wind from a neighbouring farm. A famous series of precedent-setting court cases ensued, which finally culminated in a hearing by the Supreme Court of Canada. In May 2004, it ruled in favour of Monsanto, stating that Monsanto's patent was valid and that an infringement had occurred. (For a timeline of events, see http://www.historycommons.org/timeline.jsp?timeline=seeds_tmln&seeds_legal_actions=seeds_legalMonsantoVSchmeiser.)

Food Sovereignty

The more we read about *La Via Campesina* movement of peasants around the world, the more inspired and dedicated we become to the concept of food sovereignty. Food sovereignty is the right to own all the necessary elements of our food production system. The food policy of many governments has been geared to producing high volumes of food for export. In the developing world, the World Bank and the International Monetary Fund are active in promoting the shift away from self-sufficiency by including in their loans conditions that dictate building a nation's gross domestic product. This has benefitted agribusiness rather than small-scale farmers. It has hurt the environment by encouraging chemical-intensive monocultures; it has hurt animals by introducing factory farming; and it has hurt people by encouraging dumping of over-produced foods in third-world markets. It has made possible the mass production of cheap, unhealthy foods by the processed-food industry. We believe that food self-sufficiency is a more appropriate focus for nations, bioregions, and communities. Food sovereignty is about putting the people who produce the food in the centre of the food system, and about feeding local communities rather than export markets.

When farmers spend more time on the ground and less time on tractors, they are far more connected to what is going on with their crops, and with less land to care for, they can ensure that everything grows as well as possible. Since a greater variety of crops is usually grown on small farms they can tailor the needs of the crops to soil and microclimate conditions as they vary across the landscape. Total output tends to be higher per acre on small farms due to this diversity. Finally, selling products locally—directly to consumers—can bring a higher financial return for small farmers.

Large, monoculture farms, which have only one or two crops, are much more susceptible to pest and disease problems, which can spread extremely quickly through a crop. Lack of biodiversity means fewer beneficial birds and insects to keep pests in balance, and more dependence on chemical fungicides and insecticides, thus increasing costs. Large industrial farms are also limited by market-driven commodity prices, and have little control over their income. They also incur great environmental costs. The loss of topsoil to erosion, caused by excessive machinery use and lack of biodiversity, is a major problem across North America. Synthetic pesticides, herbicides, fertilizers, and the heavy concentration of animal manure have detrimental effects on soil, water, and biological organisms, and on the people who work on those farms.

The Effects of Fertilizers and Pesticides

Nitrogen fertilizer is synthesized through a process that creates ammonia from natural gas.

Currently, 5 per cent of global natural gas consumption is used for this purpose. Ammonia is then a component in the forms of ammonium nitrate and urea, which are applied as fertilizer. This technology was developed when a huge surplus of ammonium nitrate, used for making explosives, was diverted to fertilizer production after the Second World War as a way of disposing of it. The production and application of nitrogen fertilizer make it a major greenhouse gas producer, due to the release of nitrous oxide, ammonia, and carbon dioxide. Globally, the most important natural source of nitrous oxide is known to be soil, because of the increasing concentrations of agricultural chemicals. Soil can account for 75 per cent of global nitrous oxide emissions.[5]

The highly soluble nature of fertilizer means that the runoff commonly causes eutrophication, a form of nutrient pollution, which disrupts the natural ecological balance in water systems. Algae in the surface layer proliferate, causing oxygen depletion below, and this results in the die-off of species. Various "dead zones" have been identified around the world where this process is occurring. A massive dead zone exists in the Gulf of Mexico, where the Mississippi River releases pollutants derived from agribusiness activities (primarily animal feedlot wastes and fertilizer and/or pesticide runoff) into the ocean. Its size varies, but it has covered an area up to twenty-one thousand square kilometres, or the size of New Jersey.[6]

Pesticides are a broad term for a class of chemicals that are used for the prevention, control, or decrease of pests. Pests could be insects, bacteria, weeds, fungi, rodents, nematodes, molluscs, or mites. Pesticides work by poisoning their

No chemicals on the farm means a healthy environment for raising children.

subject. Like fertilizers, they came into broad use after the Second World War, when US chemical companies shifted their wartime research to opportunities in the domestic market. Teaming up with agricultural experts, they promoted the use of synthetic chemicals as pesticides. The most famous one is DDT, which was used among troops during the war to control typhus-spreading lice in southern Europe, and malaria-spreading mosquitoes in the South Pacific. DDT came to be considered the "wonder insecticide," as it was inexpensive, effective on a wide range of insect pests, persistent in the environment, and insoluble in water, and didn't appear to be toxic to mammals. It became the most widely applied chemical in human history. We now know it to be directly toxic to fish and crabs, and indirectly toxic to other organisms, having a devastating effect as it biomagnifies up the food chain to predatory birds such as hawks and eagles. DDT was banned in the US in 1973 but continues to be produced in developing nations.[7]

The spraying of insecticides to control insects, and herbicides to control weeds, is common practice on large and small non-organic farms. However, over 98 per cent and 95 per cent of these chemicals, respectively, reach destinations other than their target species.[8] The use of pesticides has become so widespread that many of these target species, both plants and insects, develop resistance to the chemicals, creating a need for stronger chemicals. The number of insect species known to have displayed pesticide resistance has increased from fewer than twenty in 1950 to more than five hundred in 1990.[9] Pesticide drift contributes to air pollution and threatens other species. Pesticides are found to pollute groundwater, waterways, and wells through spray drift, runoff, spills, and leaching through the soil. The problem is now so widespread in the US that maximum contamination levels have been identified for pesticides in public bodies of water. Pesticides are also identified as major soil contaminants, along with petroleum products, lead, and other heavy metals.

Bees are incredibly important contributors to the plant world as they are primary pollinators, participating in the sexual reproduction and cross-pollination of plants. Cross-pollination is essential for some species, and promotes genetic diversity for others. Pesticide misuse kills bees, either through direct spray on the worker bee or causing kills back in the hive by contaminating nectar or pollen. At least fifty-nine pesticides approved for use on crops in the US are categorized as "highly toxic" to bees, with an additional twenty-seven listed as "moderately toxic."[10]

As organic farmers, we are committed to approaching the growing of food with an ecological mindset. There is no doubt that we are disrupting the land with our cultivation machines and rows of crops, and our weeding out of unwanted plant species. However, we grow or encourage the flowering plants that attract beneficial birds and insects, which in turn prey on pests. We are attentive to their zones of habitat around our fields. We constantly add organic matter to the soil, to help balance the nutrients and encourage microbial activity. We aim to replace, through natural means, the nutrients that are taken up by our crops. Ideally, our actions should leave the soil and the water table in better health than when we started. Those in our community who have been cultivating the same soil

Tree frogs are a common sight on our farms and are an indicator of the health of the environment. They are one of the first things to disappear as toxicity increases.

for a number of years are seeing these effects. Traditional forms of agriculture, like organic farming, were self-sustaining for thousands of years before the advent of synthetic fertilizers and pesticides. Traditional agriculture and aquaculture are still practised in developing countries all over the world, in a truly sustainable fashion.

The Green Revolution and Loss of Traditional Agricultural Practices

Traditional agricultural technologies are increasingly being threatened and replaced by industrial systems that fail to take into account the holistic and ecological nature of the earlier methods. The term "Green Revolution" refers to the industrialization of agriculture in developing nations between the 1940s and the 1970s. It is usually touted as having staved off starvation for millions, under the premise that traditional forms of producing food could not support the rate of population growth on the planet.

However, more people are hungry now than before the Green Revolution. This is hard to accept, because we are immersed in a culture that believes science will save us. While globally there is more than enough food produced, current distribution systems leave millions starving. The high costs of Green Revolution technology (hybrid seed, fertilizers, and pesticides) have driven small-scale farmers out of business. Food production has become the domain of large corporate farms that are part of governmental geopolitical agendas. Food prices have become disconnected from production costs, farmers have become alienated from the land, and people have become minions of big business, rushing out to buy the food products that are advertised rather than relying on their traditional foods. People are getting sick, the landscape is being ruined, and corporations are getting rich while farmers are going bankrupt.

The staggering complexities of food production and distribution systems are not adequately acknowledged in the understanding of history espoused by advocates of the Green Revolution. For example, in India, peasants grew a wide variety of legumes, fruits, and vegetables, which gave them food self-sufficiency. Their grain crops provided straw that was used as fodder for their cattle, which in turn produced milk, and manure for fuel and fertilizer. They saved their own seeds, and over time, developed thousands of varieties of crops, all locally adapted.

With the advent of industrial farming, US agencies such as the Rockefeller and Ford foundations brought their products to the developing world. They established co-operative relationships with the Indian government to promote an industrial, monoculture system of food production with promises of very high yields. As a result many farmers switched to growing wheat and maize. These were high-yielding varieties that required more nitrogen than their ancestors; they were developed to grow best with chemical fertilizers and pesticides. The plants tended to fall over under their heavy seed load, so they were bred to be dwarf in size.

Several things happened. Farmers indeed enjoyed greater yields. However, due to the smaller dwarf plants, there was much less fodder available for their farm animals, which meant less manure for fuel, resulting in the need to buy kerosene. Less manure also meant less organic matter to support soil fertility. Their crops became dependent upon chemical fertilizers and pesticides, because monocultures are not ecologically balanced and pests thrive under those conditions. Indian farmers were encouraged by their government to stop growing their food crops in favour of cash crops for export (wheat and maize). The farmers and their families had less to eat, and were reliant upon seeds and chemical inputs from an outside source, and thus were subject to price increases on those items.

As more grain was arriving on international markets, farmers were going hungry, their land was being stripped of fertility, and their wonderful genetic wealth of seed diversity was being lost. Vandana Shiva is a physicist and food activist, author of many books on food security, and winner of the Alternative Nobel Peace Prize in 1993. She writes that "... as more grain is produced and traded globally, more people go hungry in the Third World."[11] Furthermore, she claims that industrial monocultures that are maintained by intensive external inputs, such as fertilizers and pesticides, create food *in*security:

> A study comparing traditional polycultures with industrial monocultures shows that a polyculture system can produce 100 units of food from 5 units of inputs, whereas an industrial system requires 300 units of input to produce the same 100 units. The 295 units of wasted inputs could have provided 5,900 units of additional food. Thus the industrial system leads to a decline of 5,900 units of food. This is a recipe for starving people, not for feeding them.[12]

Rachel and Jade at an agricultural rally.

Social Costs of the Green Revolution

A look at the socio-agricultural perspective in the decades since the onset of the Green Revolution reveals a bleak picture: families losing ancestral land as their debt reaches crisis proportions is a story playing out all over the world. In a country like India, where farmers were already poor, this is resulting in the additional disaster of skyrocketing rates of farmer suicide.

> In Punjab, the epicentre of the country's high-tech agricultural "Green Revolution," the United Nations scandalized the government when it announced that, in 1995–96, over a third of farmers faced ruin and a crisis of existence ... This

phenomenon started during the second half of the 1980s and gathered momentum during the 1990s. It has been getting worse. According to the most recent figures, suicide rates in Punjab are soaring. As one newspaper put it, this sad end to the farmers who were meant to have thrived under India's brave new agricultural future is a "Green Revocation."[13]

The problem is not isolated to India and thus cannot blamed on inadequate or corrupt government leadership. In China, farmer suicide and attempted suicide rates through ingesting pesticides are high. "Of a sample of 882 suicides in China from 1996 to 2000 . . . agricultural labourers made up over half the dead." Farmer suicides have increased in wealthy nations as well. It was widely reported in the media in 2006 that farmer suicide rates soared to one every four days in rural Australia. Though these suicides were closely related in the media to a five-year drought, the figures show one suicide every six days a decade earlier, according to an academic study conducted from 1988 to 1997.[14] "In the UK, farming has the highest suicide rate of any profession."[15] Here in North America we are familiar with the disappearance of the family farm, and the distress and depression caused by the loss of family land due to crippling debt burdens.

Corn and the Industrialization of Farming Systems

Several of the problems associated with industrial agriculture and our industrial food supply can be illustrated by the rise of corn as a staple crop and commodity in the US Farm Belt. Michael Pollan writes a compelling chronicle of the history of corn in his book *The Omnivore's Dilemma*. We have relied heavily on his work for the pages that follow.

Beans are one of the earliest known plants cultivated by humans. These are Orca beans, an heirloom variety from Mexico. They can easily be saved and propagated by the home gardener.

Hybrid Seed

F1 is the designation given by seed breeders to the product of breeding two open-pollinated parent plants together. The parents are selected for specific traits, such as disease resistance, productivity, or appearance, to name a few. The F1 generation will not produce identical offspring, so the farmers have to return to the seed companies to purchase more the following year. Parent plants are kept a trade secret. Some farmers are regaining control over the seed supply by doing participatory plant breeding in conjunction with universities. Notably, this is occurring in Cuba, where the food supply has been a critical issue.

Catherine lovin' up the kale buds at an early spring market booth. "She could sell kale to the kale-weary."

Corn and soybeans are grown on a rotational basis, and both have become mainstays in the processed-food industry in North America. When seed breeders started developing and tailoring hybrid corn seed to suit industrial production methods, the game changed.

Hybrid F1 was the product of breeding two open-pollinated parent corn plants together. This first-generation hybrid had some unusual characteristics: genetically identical plants, and significantly higher yields than its parents. These traits alone set the stage for mechanical cultivation, and more land was allotted to corn production.

Significantly, breeders found that seeds saved from F1 hybrid cobs and replanted produced very poor results, making them useless to the farmer in subsequent years. The farmer had to return to the breeder to buy more of the first-generation seed.

shifted more and more production to this cash crop. In the 1920s yields were roughly twenty bushels per acre. By the 1950s they had risen to seventy to eighty bushels per acre, and with further tinkering, including genetic modification, yields can now reach two hundred bushels per acre. Hybrids have thicker stalks and stronger root systems than their open-pollinated ancestors, which allows them to be planted extremely closely: thirty thousand seeds per acre, as opposed to eight thousand in the 1950s.

The rise of corn could not have taken place without the concurrent rise in the application of synthetic fertilizer. Before it was made available through the use of fossil fuels, farmers were dependent upon the nitrogen that was "fixed," i.e., collected from the air and released into the soil by bacteria living on the roots of legume plants. Crops were regularly rotated with legumes to restore the nitrogen that had been consumed by the preceding crop. Corn was used partially as fodder for animals, which in turn produced manure, which was spread on the land to restore vital nutrients. Whereas farmers had previously relied on the energy of the sun for nitrogen supplies and thus soil fertility, with the advent of fertilizer, they came to rely upon fossil fuels. The biological cycle was broken, and farmers no longer needed legume rotations or animal manure.

An industrial mindset became prevalent, and monoculture the norm. This transformed North American farms. Tractors and mechanization contributed to the displacement of horses, and the other farm animals disappeared to leave more land for corn production. Much diversity of livestock, landscape, crops, and people also disappeared, giving way to our current monotonous

Though common today, the concept of buying seeds from a corporation was revolutionary in the farming world. Until the 1930s, farmers saved their own seed or traded with neighbours, and over time these seeds developed genetic characteristics that adapted them to their specific ecological circumstances. Hybrid corn, however, offered such huge yields and was so well adapted for mechanical harvesting that farmers steadily agricultural landscapes: rows and rows and rows of corn, punctuated only by roads, fences, and far-flung farmhouses.

As in India and other areas affected by the Green Revolution, farmers shifted away from sustainable, diverse production of plants that not only fed their families and their livestock but also maintained the fertility of their soil. They moved toward the production of a commodity, corn, which could not feed their families, and which bankrupted their soil fertility, causing dependence on external chemical inputs.

us Department of Agriculture Policy and its Long-term Implications

Accompanying the corn-yield breakthroughs were changes in federal farm policies, which removed the checks and balances that had safeguarded both farmers and the marketplace from wild fluctuations in supply and demand. Previously, farmers were given the option of a government loan if the supply of a storable commodity was high and prices were down. They could store their grain in a government granary until the price came back up, then sell and repay the loan. If the price stayed depressed, they would give up their grain to pay back the loan, and the government would store it for a longer period, selling it at some future time when supply was low and prices came up. This didn't cost the government much, since most loans were repaid. More importantly, it discouraged dumping by regulating the amount of product on the market, and thus the prices, keeping surpluses back and releasing more grain when times were lean.

During the Nixon administration, these and other safeguards were eroded in the interests of increasing production and driving down prices. Farmers were encouraged to grow "fencerow to fencerow" and "get big or get out." This served the interests of both corporate food processors, which were looking for cheap raw materials, and grain exporters. The federal granary was abolished and loans were replaced with direct payments: farmers were compensated for the amount of money they lost below the target price. Rather than keeping an excess of corn out of the market, this encouraged farmers to sell their corn at any price. Predictably, the price kept falling.

The target price upon which the subsidies are based has gradually been legislated down over the years and farm incomes have fallen along with the prices, forcing thousands of farmers into bankruptcy. The remainder, locked into corn and soybean production due to their machinery, infrastructure, and subsidy structure, and the insistence of grain elevators to be supplied with only those two crops, attempt to produce ever more, just to keep up with their bill and debt payments. The result is twofold: staggering amounts of corn being produced per acre, and ever lower prices. What would seem to be a success has become a spiral of financial failure for these farmers Not only does the us federal treasury now spend up to five billion dollars a year subsidizing this cheap corn, but subsidy payments make up nearly half of net farm incomes.[16]

Through subsidies, government policy is largely responsible for the loss of small farms. Now, corporate-owned farms dominate food production in the us.

In 1999, over 70 per cent of subsidies went to just two commodity crops: corn and soybeans.

These supports promote industrial-scale production, not small, diversified farms, and in fact create an environment of competition where subsidized commodity producers get help crowding the little guys out of business. It is this, rather than any improved efficiency or productiveness, that has allowed corporations to take over farming in the United States, leaving fewer than a third of farms still run by families.[17]

The beneficiaries of commodity corn are corporate agribusiness entities, which are busy finding all kinds of uses in consumer products for what has become a cheap, abundant raw material. We, the consumers, are eating this corn in its many guises, mostly in processed food: not only as corn syrup, corn oil, corn flour, and corn starch, but also under the names of glucose syrup, maltodextrin, crystalline fructose, ascorbic acid, lecithin, dextrose, lactic acid, lysine, maltose, MSG, polyols, caramel colour, and xanthan gum. Corn is used in the industrial sector as starches for strengthening and finishing. It is a component of textiles, plastics, adhesives, construction materials, pharmaceuticals, and cosmetics. Corn sweeteners are not only in food and beverages, but also in intravenous fluids, vitamins, amino acids, and alcohols. Corn is used as animal feeds and in pet foods. Most recently, an industry has developed to produce ethanol from corn.[18]

Agrofuels

There are many reasons to oppose the production of ethanol and other agrofuels. Agrofuels are made from biomass including corn, soy, palm, and sunflower oils, which are diverted from the food stream. The larger class of biofuels can include fuels made from garbage, algae, trees, and crop residues. First, agrofuels are grown using land that could be growing food. The World Bank estimates that 75 per cent of the food cost increase in 2008 was from agrofuels. The United Nations says it takes two hundred and thirty-two kilograms of corn to fill a fifty-litre car tank with ethanol. That is enough to feed a child for a year. Even when fuels are made from crop waste and grasses, biomass is diverted from fodder and from soil building, which is a net loss to food-production capacity.

Second, agrofuels are not environmentally friendly. According to studies at Cornell University and the University of California at Berkeley, they take at least 30 per cent more energy to produce than they provide. The food components are grown industrially, using nitrogen fertilizer that is a major greenhouse gas contributor, and this makes them more polluting than mining and burning petroleum.

Third, they are distracting our attention from truly ecologically sound policies and practices. Under their green-washed agenda, agribusinesses can usher in more genetic engineering (most agrofuel crops are GMO), which threatens to contaminate the natural world and the crops of other farmers around them.

Agrofuels are a distraction, not a solution. Reducing car use, and consumption in general, and focusing research on true alternatives to carbon-based fuels are some of the ways we need to proceed.

Raising chickens and other livestock on pasture makes for contented animals and more nutritious eggs and meat. The manure contributes to farm fertility, as opposed to a factory farm model where the manure becomes an unhealthy waste disposal problem.

Factory Farms

Corn-derived products have arisen more out of the need to use up what Michael Pollan calls a "huge surplus biomass" than to meet any true requirement. Industrial animal feedlots developed as one of the ways to absorb these massive amounts of corn. Virtually all of our chicken, beef, and pork now comes from factory farms where tens of thousands of animals are crammed together and fattened up as quickly as possible. Pigs and chickens are adapted to eating corn, but cattle are particularly inefficient at converting corn into meat, having evolved as grass eaters. They have to be introduced to the corn diet gradually and they suffer from a near-constant state of bloat and indigestion. For this reason, feedlots employ resident veterinarians and cowboys whose sole job is to patrol the pens looking for cows in distress. Crowded into small pens, the animals stand in

their own excrement from the time they arrive at the feedlot for "finishing." They are transported in inhumane conditions, and slaughtered en masse in production-line facilities that, like the rest of our industrial world, are more about output per hour than the health and well-being of the animals. Three key issues associated with factory farms fly in the face of sustainable agriculture: cruelty to animals, overuse of antibiotics, and manure disposal.

Animal Welfare

Laying hens live in tiny cages, stacked floor to ceiling, in which they cannot stretch out a wing. Their beaks are clipped early on to minimize the effects of stress-induced pecking of their cage-mates. When their egg production begins to slow they are "force moulted": starved of food, water, and light to stimulate one more bout of egg-laying before they are killed.

Broiler birds are packed together into huge warehouses, and never see the light of day. The ammonia rising from the vast amounts of waste building up underneath them causes burns to their skin, eyes, and respiratory tracts. They are bred for large thighs and breasts, and for rapid weight gain, which can cause spontaneous heart attacks as their organs cannot keep up with their body weight. Deformed and broken legs are common for the same reason.

Sows are kept in cages so small that they cannot even turn around. Piglets are weaned at ten days because they grow faster on hormone- and antibiotic-fortified feed. However, the need to suck doesn't leave them, causing them to chew on the tail of the pig in front of them. After some time in close confinement, without seeing natural light, unable to exercise, roll in mud, root, or breathe fresh air, pigs become demoralized, and will allow other pigs to chew their tails to the point of infection. The USDA recommends tail-docking as a means to treat this problem. The tail is cut off only partially, to a stub, rendering it more sensitive, so the animal will be in more pain and will fight back to defend itself.

Veal calves are kept in solitary confinement. Movement is discouraged so their muscles stay soft and the flesh remains tender. They are deprived of light for most of the four months of their lives, and are fed a diet deficient in iron to keep their flesh pale and appealing to consumers.

Antibiotics

Antibiotics are a crucial component of industrial meat production. The animals are raised in conditions for which they have not evolved; they are subject to unnatural diets and intense crowding, which cause stress and prohibit normal socialization. These conditions leave the animals prone to disease. Antibiotics are used not only for medical purposes; they are routinely added to feed to speed weight gain. This translates into greater profits for the industry. According to some estimates, 70 per cent of the US production of antibiotics goes to livestock. This overuse leads to the rapid evolution of "superbugs" or antibiotic-resistant strains of bacteria. Through runoff, antibiotics from urine and manure accumulate in waterways, disrupting natural bacterial balances and eventually turning up in drinking-water supplies.

Manure

Wendell Berry once wrote that when we took animals off farms and put them onto feedlots, we had, in effect, taken an old solution—the one where crops feed animals and animal waste feeds crops—and neatly divided it into two new problems: a fertility problem on the farm and

a pollution problem on the feedlot. Rather than return to that elegant solution, however, industrial agriculture came up with a technological fix for the first problem—chemical fertilizers on the farm. As yet, there is no good fix for the second problem.[19]

One of the most beautiful things about mixed organic farms is that there is very little waste: manure is a valuable resource for soil fertility. In the concentrations produced by the industrial system, however, it is a pollutant. Organic farmers cannot use this manure because it is contaminated by antibiotics and genetically modified feed grains. Compared to chemical fertilizers, manure is bulky and inconvenient, so conventional farmers don't use all of the excess either. Tragically, it sits unused, leaching nitrates into waterways, contributing to eutrophication, and causing disease outbreaks. Human health is put at risk, as in September 2006, when two hundred people in North America were made dangerously ill and three people died from eating spinach irrigated with groundwater contaminated by E. coli bacteria from feedlot manure.

> The lethal strain of E. coli known as 0157:H7, responsible for this latest outbreak of food poisoning, was unknown before 1982; it is believed to have evolved in the gut of feedlot cattle. These are animals that stand around in their manure all day long, eating a diet of grain that happens to turn a cow's rumen into an ideal habitat for E. coli 0157:H7. (The bug can't survive long in cattle living on grass.)[20]

It is entirely possible to avoid these problems of antibiotic overuse and manure concentrations by raising livestock on pasture for their whole lives. The only reason we don't is that in the current system of subsidized, high-input corn production, it is cheaper to finish them in feedlots. The current industrial food system distorts the true cost of production. Our subsidized system, which props up large agribusiness, actually leads to environmental degradation. If artificially cheap grain was not available, or if farmers had to clean up nitrate pollution or pay for their carbon emissions, more sustainable production methods would be economically viable.

Reducing a food-production system to an industrial model costs us greatly in terms of health and environmental damage. The only apparent benefit is cheap food. Ironically, even this "benefit" is costing us, since the cheapest food to come from this system is the unhealthy, processed food that is leading to so many of the health issues that are clogging our arteries and our health-care system. When we take a broad view, the illusion of the low cost of food is exposed.

E. coli and the Fear of Food

Ironically, in this age of industrial, highly regulated food production that is meant to be clean, safe, and sterile, we are the first generation of eaters to fear our food. There is rising concern in North America about basic food safety. We hear about E. coli on spinach and listeria in processed meats. These outbreaks demonstrate failures in the industrial machine. Minor errors are magnified, because our food system is so centralized. Outbreaks are particularly frightening to the consumer because they highlight how little control and understanding we have of the system that feeds us.

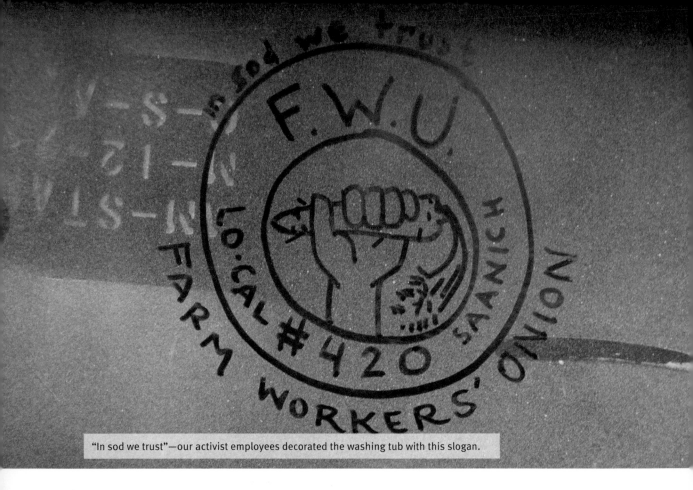

"In sod we trust"—our activist employees decorated the washing tub with this slogan.

Farm Workers

It is not only the environment that suffers from policies that keep food as cheap as possible. Poor labour standards are another consequence of this system. Even here on Vancouver Island, farmers bring workers up from Mexico because they are unwilling provide working conditions that would appeal to the local labour force. Chemical use on farms, and the lack of safety equipment, has led to myriad health problems in farm workers. They are also often put in physical danger. In March 2007, in the Fraser Valley of BC, an overcrowded van carrying farm workers rolled over. Three workers were killed and fourteen were injured. The van was designed to hold ten people but was transporting seventeen, having had its seats removed and replaced with unsafe wooden benches. This accident could have happened to any vehicle, but because there were no seatbelts, people lost their lives. This is not an isolated incident. The pressure on farmers to keep prices low encourages them to cut corners wherever they can.

Moving Forward

The next time you are in the grocery store, take a minute to think about the price of an "affordable" produce item. Consider what's behind it: the grocery store's markup, the distributor's markup, packaging, refrigerated transportation, water, machinery, diesel, seed, fertilizer, and land costs. How much is left to pay for all the labour that went into seeding, weeding, harvesting, and packing it? Unfortunately, a higher price in the grocery store doesn't necessarily mean better labour standards. The only way to know for sure is to know your farmer.

We didn't know this complex reality when we started farming. We had a vague notion that something was wrong, but we hadn't connected the dots. When you do something well, you have a sense of its true value. After working hard to grow a bed of beautiful lettuce, we have no hesitation in charging $3.00 per head at the market. Seeing an imported head in the grocery store for $0.99 brings home the reality of the corners that must have been cut. The longer we farm, the more knowledgeable and passionate we become about the issues behind our food system. It's impossible for us not to be advocates, activists, and educators. At the same time, our first goal is finding a better way to grow more and more food to change our little corner of the system.

Jeremy and Larkin flex some muscle in the field. We have been blessed with farmhands who are hard-working and a lot of fun.

Shopping

We have each had moments when we looked around the grocery store and realized that for most people shopping is a much simpler affair than it is for us. No longer can we just pick up items that look appealing or inexpensive. Now there is so much more to consider: Is it organic? If so, are *all* the ingredients organic, or just a few? How far away was it grown? Were the workers treated fairly? Is it healthy? Is it in season locally? How is it packaged? What are the politics of the company that produced it?

Sometimes we are envious of people who don't face questions of conscience with every purchasing decision.

In 2009, the majority of our farmhands identified as gay and queer. So we joined in the fun by sponsoring our own entry into the Pride Parade, "The Rainbow Chards," complete with garlic whips and mud-spanking.

Endnotes for Chapter Four

1 http://www.seedalliance.org/index. php?page=SeminisMonsanto

2 Pimentel *et al*, 2005, "Environmental, Energetic, and Economic Comparisons of Organic and Conventional Farming Systems," *Bioscience*, Volume 55, pp. 573–582

3 Peter Rosset, 2003, "The Multiple Functions and Benefits of Small Farm Agriculture in the Context of Global Trade Negotiations, *The Society for International Development*, 43(2), pp. 77–82

4 Utviklingsfondet/The Development Fund, Norway, 2010, "A Viable Food Future," retrieved June 7, 2011, from www.utviklings-fondet.no/viablefoodfuture/

5 J.J. Yienger and H. Levy II, 1995, "Empirical model of global soil-biogenic Nox emissions," *Journal of Geophysical Research*, Volume 100, pp. 11447–11464

6   *National Geographic*, 2008, "Dead Zone," encyclopedia entry, retrieved from *education. nationalgeographic.com/encyclopedia/ dead-zone/*

7   Patricia Muir, 2007, "A History of Pesticide Use," retrieved from http://people.oregon-state.edu/~muirp/pesthist.htm

8   G.T. Miller, 2004, *Sustaining the Earth*, 6th edition, Thompson Learning, Inc., Pacific Grove, California, pp. 211–216

9   Leo Horrigan *et al*, 2002, "How Sustainable Agriculture can Address the Environmental and Human Health Harms of Industrial Agriculture," *Environmental Health Perspectives*, May, p. 447

10 Matt Wells, "Vanishing bees threaten US crops," BBC News, Florida, retrieved from http://news.bbc.co.uk/2/hi/6438373.stm

11 Vandana Shiva, 2000, *Stolen Harvest: The Hijacking of the Global Food Supply*, Cambridge, MA: South End Press, p. 13

12 *Ibid.*

13 Raj Patel, 2007, *Stuffed and Starved: Markets, Power, and the Hidden Battle for the World's Food System*, Toronto: HarperCollins, p. 25

14 ABC Television Media Watch, "The Decade Old Headline," February 2007, retrieved from http://www.abc.net.au/mediawatch/ transcripts/s1869891.htm

15 Patel, *op. cit.*, p. 28

16 Michael Pollan, 2006, *The Omnivore's Dilemma: A Natural History of Four Meals*. New York: Penguin, p. 24

17 Steven L. Hopp, 2007, in Barbara Kingsolver, *Animal, Vegetable, Miracle: A Year of Food Life*, Toronto: HarperCollins, p. 206

18 Pollan, *op. cit.*, p. 37

19 Pollan, *op. cit.*, p. 54

20 Pollan, 2006, "The Vegetable-Industrial Complex," *The New York Times*, October 15

CHAPTER FIVE

Coming Together as Saanich Organics

Saanich Organics is a co-operative vegetable-marketing business that operates in the Victoria region, an epicentre in Canada for small farms that are selling directly to their communities. Saanich Organics works for three reasons: first, we, the owners, are compatible and complementary; second, geographically we are close to each other and to a booming market; and third, our combined resources have enabled us to gain efficiencies of scale and to be resilient in a very insecure niche.

We are fortunate to have a physical location that contributes to the success of our business. Our mild climate is conducive to the year-round growing of diverse crops. We live in a wealthy and progressive area. Residential customers are drawn to our produce for a variety of reasons: environmental commitment, health, and politics. Our chef customers also have various reasons for buying from us: some are capitalizing on the trend of culinary tourism, others share our commitment to building a sustainable local food system. All of our customers value the superior taste of our fresh, well-grown produce.

Saanich Organics has grown in the secure and sheltered environment of Northbrook Farm in the Mount Newton Valley near Victoria. Access to land is a major issue for new farmers in this region. Land prices are high and leases can end abruptly. It has been a boon to be able to build a greenhouse, invest in irrigation and soil amendments, and be confident that we won't lose those investments. We have the additional benefit of low-rent access to a garage that serves as our box-packing facility.

You may not have such a progressive population around you, but take stock of what could work for you. Know that the organic market is growing and that there are opportunities for aspiring farmers. Unlike us, you may have the great advantage of affordable, fertile land. However, if your geographic location isn't ideal, it may be even more important for you to co-operate with other farms around you.

Background

Our friends and mentors, Tina Baynes and Rebecca Jehn, began Saanich Organics in 1993 in order to sell their produce directly to customers through a weekly vegetable home-delivery service. When they sold us the business in 2002, not only did we purchase a box program, we also inherited a business that was run with the utmost of integrity. Their customers received only the highest quality vegetables, grown and harvested by two women who were meticulous and passionate about healthy food. We had big shoes to fill and felt a combination of excitement and trepidation about running a business with such an admirable history. From the box customers' point of view, little has changed. We still run the box program with the basic structure that Tina and Rebecca established. The business has now grown to include sales at farmers' markets and to restaurants and grocery stores, as well as L.J., our experiment in co-operative greenhouse-crop production. (See Rachel's diagram in Appendix C.)

Saanich Organics included a combination of physical assets and goodwill. The physical assets were sixty wooden boxes, a commercial salad spinner, a few Rubbermaid bins, and plastic bags. More important by far was the intellectual property, including the customer list, the business's good name and reputation, and the administrative system. We hired an administrator to run the business side, and she computerized the system that has since undergone many incarnations.

Now that we have a professional bookkeeper to keep us on track, we laugh at how loose we allowed the finances to be in the early years. At that time, we really never knew if the ink was black or red. As long as our administrator said there was enough money in the account to pay the farmers, we didn't ask questions. Years ago, Rachel and Robin realized that if they didn't set financial goals for their own farms, they wouldn't achieve them. Similarly, Saanich Organics didn't make money until we turned our focus to our finances. We now have monthly financial meetings where we review detailed accounts of each part of the business. While Saanich Organics now reliably produces a small profit, it is not the bulk of our income; it functions primarily as a vehicle for selling our produce efficiently, thus making our farms more profitable.

Saanich Organics' mission is to make the small organic farms in our region financially viable by efficiently selling high-quality produce at prices that reflect the cost of production. Forming our own distribution network has given us more control over the market and over the prices we receive for our produce. Selling solely to a for-profit distributor wouldn't provide us with a viable return. By welcoming other farmers to sell through our network, we handle a volume of produce that allows us to hire out the chores of marketing and delivering. We are able to sustain our low markup and take on additional employees at a fair wage to further lighten our workload. We have the option to free up more time for farming, or take on paid positions within the Saanich Organics structure. By cooperating, we achieve greater efficiency, increased production, cost savings, and a better quality of life.

Life Choices

Before addressing the reasons behind Saanich Organics, it seems logical to address the reasons behind our farming. None of the three of us had any background in agriculture, nor any formal farming training. Yet stubbornly, against the grain, we started our farms from bare pasture and never looked back. This life choice was a complicated synthesis of politics, practicality, and soul searching. It stemmed from a desire to have a light environmental footprint, a fear of having a desk job, and a search for a meaningful life balance. Since then, our passion for food production has broadened to notions of social justice through food sovereignty, and concern for the plight of farmers around the world.

The timing was exactly right for our entry into the field. The market for organic produce was just taking off, and the momentum behind the sustainable-agriculture movement was captivating. Leaders in the organic movement were making compelling arguments against the domination of conventional agriculture and the multinational corporations that fuel it. This left a lot of room for new farmers to re-create agriculture that suited their bioregions, their scale, their values, and their customers. Times were changing and we were riding on the cusp. We were physically exhausted and sleep deprived, but mentally engaged and grinning from ear to ear.

The mainstream media often tell hard-luck stories about agriculture; articles about drought, mad cow disease, and chronic financial losses lead one to think that farming is hopeless. This bleak outlook is real for many farmers, but it is not our experience. A climate that allows year-round growing, combined with open-hearted farmers and educated, health-conscious consumers make this area a perfect breeding ground for the organic movement.

In a shift back to a sustainable-farming model, organic farmers are encouraged to use our common sense and to rely on our community, rather than calling in experts for advice. Whereas specialization was once a virtue, organic farmers delight in diversity and in the complexity of natural systems. Whereas physical labour was seen as drudgery, organic farmers enjoy the physical nature of the work—being outdoors, eating with the seasons,

Japanese turnips are a staple for us nearly all year long. They grow quickly and well, and they taste delicious!

Karen and Heather pack boxes on a cold winter night.

and feeling connected to the land. Whereas farmers used to be seen in a derogatory light, organic farmers are emerging as heroes and being positively portrayed in the media.

It wasn't all bliss by a long shot. Seven-day workweeks all summer, balancing farming with other jobs, crop losses, and not enough infrastructure on our farms caused stress and despair. The desire to succeed added a layer of anxiety to the practical challenges, and the urge to save face was strong in light of all the family members, friends, neighbours, and strangers who said it couldn't be done. Financially, it seemed impossible in the beginning. Robin didn't even keep records for her first two years because she didn't want to know about all the money she wasn't making; Heather calculated her wage at about $5.00 an hour, if she didn't include expenses.

How We Came Together

The three of us met through farm community gatherings, and we were all former suppliers of Saanich Organics under the previous owners.

We had attended work parties together and were all vendors at the Moss Street Community Market. But the notion of working together had never occurred to us until Rebecca and Tina approached us offering to sell Saanich Organics. At that point, Saanich Organics operated only the box program, the vegetable home-delivery service that went to twenty families once a week.

Rachel was the driving force behind the purchase because she recognized the value of having control over the market as a distributor rather than just being a contract grower. She felt that the box program should thrive, given the healthy state of the local organic food movement. Robin was wary of taking on the tedium of administering the box program because she had seen firsthand how much time Rebecca and Tina spent on the ledgers and maintaining customer lists. Heather felt that working together would be more work than farming alone. However, when faced with the prospect of being at the mercy of middlemen, taking on paperwork didn't seem so bad. We all agreed that Saanich Organics filled a valuable need, both for the farmers and for the customers.

We wanted to honour Tina and Rebecca's vision and carry on the good work they had started.

The market for organic produce had increased tremendously in the nine years Saanich Organics had been in operation, and this gave us flexibility in restructuring the business. First, we could add a reasonable markup to the produce without fear of losing customers. Second, we didn't have to compete with each other for precious restaurant contracts because we recognized that the market was now big enough for all the produce we could grow.

The first decision we made together was to hire an administrator, Brenda, and our trusty delivery guy, Tim. The many hours we had seen Tina and Rebecca put into the box program didn't appeal to us, so we thought we'd hire Brenda to take care of all of that. We naively thought it would be no more work than before: we'd tell Brenda what we had available for the box, and she would do the rest. We look back on that conversation and laugh. Brenda went above and beyond the call of duty, but we were so scattered that she could never really get us on track. It became apparent (when Brenda announced her resignation) that we had to be more hands-on. Although the administration does not provide a full-time job, there are endless little details to track. Much of the work can't be taken care of during regular office hours, so the administrator can end up feeling that she's always on call. When Brenda left, we hired a professional bookkeeper and started paying Heather to do some of the administration.

One of the major problems that plagued Tina and Rebecca was lack of produce to sell, especially in low season. Delegating marketing and delivering meant that we could spend more uninterrupted time on the land, becoming better growers and increasing the output of our farms.

Timing is everything in farming. During high season, each additional hour spent in the field can translate into substantial increases in production. The increase in volume and diversity of produce has, in turn, allowed us to attract and retain more customers, and to provide them with better service.

We had been running the box program for a few months when Heather's husband, Lamont, was approached at the market by a chef looking for produce. Lamont proposed the idea of collectively supplying a restaurant. Since we all met weekly with our harvests for the box program, it wasn't a big leap to begin harvesting extra produce for a restaurant. Soon, all the growers were "sharing" our restaurant contracts, as we recognized it was truly a win-win situation. From the farmers' perspective, there was less pressure to come through on an order since others could fill in if there was a shortage. For their part, the chefs enjoyed the greater selection and reliability. Our restaurant sales grew rapidly, and soon comprised a third of our business.

Being able to count on regular demand for larger volumes changed the way we worked. Rather than farming tiny patches of assorted crops, we were able to think in terms of our hundred-foot beds. We used to wait for each beet or carrot to be sold before we replanted a patch, but we can now clear out entire beds of crops as soon as they are ready, and optimize succession planting to ensure we make the most of the growing season. The larger volume of produce sold brought greater returns and let us hire field staff, and having extra hands on the farm gave us another quantum leap in terms of our ability to maximize production.

The residential box-delivery program is made more efficient and less onerous by Saanich Organics' co-operative structure. Many of our residential customers are in the box program because they want contact with their farmer. We appreciate this contact, but it can be a challenge to maintain during the busy growing season. When Saanich Organics purchased a cell phone, the administrator could take calls in the field. Finally, box customers could call us and actually talk to someone in person instead of leaving a message. We publish a weekly newsletter for our box program and this results in a stronger farmer-customer connection. When time is short we can call on each other for help with administrative details.

Working together also lends itself to information-sharing and emotional support. Local expertise on organic vegetable production is still quite limited. While the community is very supportive, and open to sharing what they know, having an inner circle of colleagues who are skilled professionals is invaluable. When we're packing up produce, there can be six to ten farmers in the room. We often discuss our challenges or successes in the field and offer each other suggestions and advice. When one person has learned a new source of seed, or has tried a different technique, she will tell us about it. This openness and desire for collective success is an important benefit of co-operation. We inherited this atmosphere of openness from our mentors, and strive to maintain it and pass it on to the younger farmers who have joined us.

Each of us is personally ambitious and is proud of successes on her own farm, but this is different from being competitive. There is plenty of ambition at Saanich Organics, and farmers delight in wowing each other when we meet to pack produce orders, but there is a sincere interest in the well-being of all other farmers in the com-munity. Our successes benefit all of us, so we each feel satisfaction and joy in seeing the others' bountiful, top-quality vegetables. Each of us knows that the quality of our produce will reflect on the others, and this sense of responsibility drives our quality up.

Farming can be very trying, and with high season comes long hours. Seasonal challenges, like spring slugs, unpredictable frosts, or excessive rain, can add to the stress on families as they try to cope with the workload. Mid-summer is especially hard because, in addition to the regular growing and harvesting schedule, the timing is tight to get winter crops in the ground to ensure fall and winter harvests. While this is not as frequent as in the early years, farmers can show up at packing time in tears and near the end of their ropes. Having an understanding support network so readily available has been a lifeline, and a stabilizing force within the group. Supporters and the supported change roles, creating closeness and soliciting opportunities for emotional release and problem solving.

The Box Program

The aim of the box program is to get fresh produce directly to residential customers with minimal packaging and effort, and to provide stable demand and income for farmers. Customers who subscribe to the program pay for ten boxes at a time. They can choose a delivery either every week or every second week. The box contains six to ten produce items, with a value of $26.50 plus a $3.50 delivery charge. We run the box program for eleven months of the year, taking a break from Christmas until the end of January.

Sometimes it is a challenge to grow the diversity we need for our box customers, but we love the feeling of packing up boxes that we know our customers will be excited to open.

Heather enjoying tastes served at the Island Chefs' Collaborative spring Food Fest. We love attending events like this where we can promote our business and also enjoy a fun afternoon.

This method of sales was inspired by Community Supported Agriculture, although Saanich Organics doesn't operate as a CSA for two reasons: flexibility and finances. Our box customers want flexibility because some of them travel and grow their own gardens. Farmers want the flexibility to experiment with growing some high-value crops in small quantities, not all of which would be suitable for box customers. For example, pea shoots are delicious garnishes for restaurants and are quite lucrative in small patches. Salsify, quinoa, and okra are not on your average family's menu, but may appeal to chefs, so allotting field space to crops like these is a smart financial strategy.

The three-part marketing format (box program, restaurant sales, farmers' markets) rewards experimentation, because chefs are very willing to buy out-of-the-ordinary items at high prices, and then farmers' market and box customers benefit from the proven winners that are grown in larger quantities the following year. Eggplant and peppers used to be rare treats, but now they are regular items in our boxes. Having different markets fosters and rewards diversity, whereas the classic CSA format fosters stability.

Our box contents vary widely with the seasons. We used to worry about filling winter boxes, but now they are bursting with root crops, hardy greens, greenhouse-grown salad, stored apples, potatoes, and squash. We add variety by including produce that we freeze in the summer months, such as berries, tomatoes, beans, and shredded zucchini. Early spring is our most challenging season because the winter crops are mostly sold and the summer crops are just being planted. At this time we rely on quick-growing crops like radishes, turnips, spinach, and the first velvety chard. Before long, the wonderful diversity of the summer crops kicks in and the boxes overflow with colourful summer favourites. The challenge of fall is choosing between the seemingly endless possibilities for each box.

The box program has grown in popularity over the years. It used to be a big challenge to get a customer to sign up for a weekly box of seasonal vegetables; the cost, the inability to choose what comes in the box, and unfamiliarity with certain vegetables were all barriers to joining. Now a lengthy wait-list for the program develops over the winter months until we can take on more customers each spring. As news of the environmental and social benefits of eating locally builds, so does interest in our food. The customer base used to be rather transient as people tried out the box for novelty but customers are now staying on the program longer. Currently, there are ninety subscribers to our delivery service.

The box program requires more administration and labour per unit of produce sold than restaurant sales but we consider it our bread and butter because it is a consistent market for a large volume of produce. Chefs can be fickle, and produce managers irregular, but the box customers are a dependable counterbalance to the rest of the business.

The Commercial Division

The name "commercial division" started as a joke because it was a single box of produce that went to one restaurant. Little did we know what was to come: thirty-five restaurants, a contract with a major grocery store chain, and two weekly sales. There is a group of chefs in this region called the Island Chefs' Collaborative (ICC), who are very supportive of farmers because they believe in the value of local production. They are also building the culinary tourism trend on southern Vancouver Island. The chefs feature local producers on their menus, give out micro-loans to new farmers, and share the names of dependable suppliers. It is largely for this reason that the Saanich Organics list of restaurant contacts continues to grow.

Dealing with restaurants has its pros and cons. The pros are that creative chefs can often use unusual items in small amounts, they will buy in bulk so that the farmer doesn't have to weigh and bunch individual items, and they can often use last-minute surplus items. Restaurants tend to be less price sensitive than retailers. The drawback is that chefs can be inconsistent. They are often very enthusiastic the first couple of times a crop is offered but their interest can wane before it has finished producing. Furthermore, their business can fluctuate from week to week,

Community Supported Agriculture

A CSA program is a set contract between customer shareholders and farmers. Customers must pay ahead and commit to taking boxes every week. Farmers divide their harvest among their shareholders. Ours is not a classic CSA because Saanich Organics allows flexible payment and delivery schedules. We charge customers for what they receive rather than simply dividing the whole season's harvest among the shareholders.

Chrystal adds her onions and carrots to a wax box bound for a restaurant. Each box has a clipboard beside it with the invoice attached. As farmers add new items, they check them off on the invoice.

leaving us in the lurch if it happens to be slow. Chefs leave restaurants frequently with no notice, so consistent orders can end abruptly. On the flip side, often those chefs will turn up at a new restaurant, bringing us a new customer.

We have buffered ourselves from this shock by taking on a roster of thirty-five or more restaurants; it is important to keep many customers on the list to deal with fluctuations in volume. It takes a while for new chefs to wrap their heads around the idea that we do not have a set product list, nor an infinite supply of each item. They are accustomed to dealing with suppliers who are pulling from warehouses rather than from fields but once they get used to our system and appreciate the quality of our produce, they often become the best advocates for our business.

During high season (May through October), the commercial division has two sales each week: Tuesdays and Fridays. For the Tuesday sales, farmers call in with their lists of available produce on Friday. For the Friday sales, farmers call in on Tuesday. When making an offering, the farmer goes out to the field and estimates the quantities of crops that will be ready for harvest, and most importantly, what they physically will have time to harvest in the forty-eight hours prior to the sale. Estimating field contents is both an art and a science. Rachel tends to be the most careful and exact of the three of us. She likes to physically look at each crop. Heather is more likely to look at what she offered the previous week and adjust up or down depending on her impression of how the crops are growing. Robin knows that others tend to be more conservative than she is, so when her farmhand does the estimates, she simply adds 30 per cent to push them to harvest more.

After a morning on the phone, and a long day in the field harvesting her own produce, Chrystal would oversee the Saanich Organics packing room and write out restaurant invoices.

Chrystal and Ilya

Fresh out of an environmental science program, Chrystal came to apprentice at Heather's farm six years ago with her partner, Ilya. After completing her apprenticeship, she stayed on, first as a farmhand, then as an administrator, and now as a farmer. In many ways, we see Chrystal as an unofficial fourth partner in Saanich Organics. Though she doesn't participate in the financial meetings, she's involved in every other aspect of the business, and she's always there to lighten the load. Her eye for detail, her sense of the market, and her high standards have helped shape our reputation. While it smarts a little bit when an apprentice excels far beyond what you yourself can accomplish, it really is the best possible outcome. Chrystal and Ilya were nurtured within Saanich Organics, with mentorship, shared resources, and examples of success. Their achievement has everything to do with their own intellect, hard work, and dedication; however, we take a lot of heart from their success because they give us confidence in our mentorship efforts.

The administrator compiles the offerings into a spreadsheet (see Sample Sales Spreadsheet in Appendix C) and then calls the restaurants to take the chefs' orders. Once the sales are complete, she divides up the orders as fairly as possible between farmers and calls each farmer with his or her harvest list. The list itemizes each crop, with the amounts intended for each restaurant or grocery store, or the box program. The sales calls to restaurants and the planning of the boxes are done at the same time. This allows the administrator to read weekly trends in restaurant purchasing early in her calls and route the produce accordingly so that everything gets sold.

The farmer harvests, washes, and bunches the produce as required and delivers it to the "boxing room" (Heather's garage) by 4:30 PM on Mondays and Thursdays. Clipboards, each one with a restaurant's packing list, are placed in a specific order around the room. When each farmer arrives, he or she puts his or her produce in boxes and labels them according to their des-

Tim does more than just deliver our produce: he's a mastermind of our produce-handling logistics, not to mention a strong contender in the zucchini bake-off.

tinations. We reuse wax-coated cardboard boxes that we pick up from other organic retailers; this saves money and reduces waste. The wax treatment makes the boxes waterproof and they are the standard in vegetable packing. We avoid wax boxes from conventional sources because they can have chemical residues. Once the specified item is packed, the farmer checks it off on the packing list. When the box packing is done, the boxes are numbered (1 of 4, 2 of 4, et cetera) to ensure that Tim, the delivery person, can track each box. The box-packing room gets very chaotic during the height of the season. There are often ten farmers and farmhands, along with kids underfoot and dogs trying to weasel their way in. The more organizing each farmer can do before arriving, the better.

Before the day is done, farmers check their invoices against the master list so that any necessary changes can be made, and totals are recalculated to account for shortages, substitutions, or last-minute sales. Saanich Organics' core growers are paid once a month, while occasional suppliers are paid each time they deliver. Despite our best intentions, every week there are discrepancies between what was ordered and what was harvested. By the end of Monday night, often as late as 10:00 PM, we are all exhausted and just want to get home to bed. However, if we don't take the time to double-check everything right away, details get missed, invoices are inaccurate, and painful forensic accounting follows.

Sales days are a lot of work, but we are moving a lot of produce. Each sales day in the height of the season, from July to October, we are packing up nearly $5,000 worth of produce in commercial sales alone (the box sales are an additional $1,500). Tim has had to remind us repeatedly about the

The Moss Street Community Market, with its policy that vendors are also the producers, ensures a fair venue for farmers, and a wonderful community spirit.

basic rules of produce-boxing. We often lose sight of the fact that he has to use every square inch of his cube van, and that poorly packed produce can be damaged before reaching its destination. In the middle of a day when he has to make up to seventy stops, one flat of spilled berries or smooshed tomatoes can push him right over the edge (and rightly so!).

The Moss Street Community Market

The Moss Street Community Market takes place every Saturday from April to November on the grounds of a school in an affluent neighbourhood of Victoria. Each of us had attended the market

as individual vendors, and it was an important step in our co-operation when we began sharing a market table.

Attending the farmers' market is a mixed blessing: we get premium retail prices and a valuable chance to interact with customers and other farmers, but preparing for the market is an onerous task, and attending every week is a large time commitment and output of energy. Leaving the farm for a whole day at the height of the growing season is difficult. Now we take turns at the market. Each of us attends every third week, and sells the produce from all three farms. Certainly this makes market day more complicated. There's a lot of produce to cram into one market space, and the farmer must

Rachel and Catherine show off the beautiful array at the market. We have put in many hours over the years developing our three-tiered display infrastructure so we can dazzle passersby.

track inventory and divide the money at the end of the day, but it is well worth the effort.

Over the years, the volume of customers at the market has increased tremendously, and the amount each customer is buying has also increased. It seems that while previously people bought one or two items, now they are shopping for their weekly groceries. It used to be that the market was barely worthwhile when time and effort were factored in, but it now rivals the box program for volume of sales. During peak season, the farmers deliver two or three pickup loads to the market for the four-hour sale. There are up to three people behind our stand and often an extra person out front restocking the rapidly disappearing produce. There have been days,

especially during strawberry season, when the lineup has been thirty people long.

We think this growth is due not only to the growing awareness of the importance of buying local food, but also to the admirable philosophy of our market. All the produce vendors are organic and all the vendors can sell only what they themselves have produced. There is a spirit of co-operation between vendors to have prices reflect the real value of food. Farmers don't dump produce, and out of respect for our livelihoods as farmers, no one tries to undercut another's prices. The physical layout of the market works well; it has a nice aesthetic, with a playground in the middle and musical entertainment. The customers appreciate both the atmosphere and the authenticity that comes from buying directly from the producers. Tens of thousands of dollars change hands every week, creating a vibrant local economy. If anyone wants to start a successful market, Moss Street is an excellent case study.

All the vendors at Moss Street are successful, but we feel we have an edge because of our sense of aesthetics. The more colours, shapes, and sizes the better. Group like items together; this will prevent things from being overlooked. Bounty sells, so load up your table! The more you bring to the market, the more you'll sell, because people like to choose from a big pile. You want people's eyes to be able to flow easily from one end of the stand to the other, to give them an overall sense of all that is there. Use the vertical space: we have three layers of display to keep everything within sight and easy reach. We've noticed that there are "magic spots" on the table where things seem to sell more quickly. Be attentive, and rotate things that aren't being seen to those spots. Everything you handle will be noticed, so pick things up and move them around. Don't sit down. Above all, allow people to see your natural enthusiasm for your product: give samples and cooking tips; get to know your regular customers' names; have your farm name clearly displayed; bring a photo album or "brag book." Anything that connects customers to your farm is indispensable. Try to get the same spot every week so customers will know where to find you, and attend the market regularly so they know they can depend on you.

In 2008 we tried a second market, which was in our rural area. We were excited about selling closer to home, but we were disappointed in the volume of sales. There was a big cultural difference between this market and Moss Street.

Customers were looking for cheap produce and the farmers were only too willing to oblige. The clientele was much older, and they were less passionate about the produce. We had noticed all these qualities about this market previously but we thought that we could change it with our presence by either attracting our usual clientele or by changing the clientele that was there. In retrospect, this was naive. We've since joined another farmers' market in Victoria that is proving to be lucrative. This experience taught us that there are different types of farmers' markets, so when you're deciding where to sell, invest some time in visiting the markets in your area before making your decision.

Long John: The Greenhouse

By the end 2003, our second year of running Saanich Organics, we had a vague notion that the business had made some money, and that there was potential in our working together to grow and sell more, and to achieve greater financial success than we could on our own. The main challenge we each faced was building our own farms to the point of financial sustainability. A related challenge was having a sufficient selection of desirable crops, especially in the winter, to keep clients interested in regular, year-round delivery. The investment of Saanich Organics' profit (with a further personal investment from each of us) in the purchase of a hundred-and-sixty- by twenty-foot greenhouse was an attempt at solving those problems.

The greenhouse was erected on Heather's land, and the long, thin structure quickly earned the nickname "Long John" or "L.J." After the six-month epic of erecting the greenhouse (see Robin's chapter), we were in the salad greens business. In the first year, the three of us did a lot of work unpaid, but we hired a woman to do the harvesting. Volunteer work makes the books look good, but we were exhausted; even the financial success was compromised because we didn't have the time or energy to maintain the crops to their optimum potential. We got the greenhouse seeded during work parties, but never managed to stay on top of the weeding. This meant that our employee, Alex, spent much of her time sorting through the greens she had cut, picking out weeds. The situation got worse week after week, but because our own farms were our first priority, it was difficult to find the time to till in beds and replant them on a consistent schedule. Lack of attention is costly, and this was enough to convince us that we couldn't do it on our own. The following year, we hired more people to work in the greenhouse and paid Rachel to manage the project.

This pattern of experimentation has repeated itself with each division of the company. For one or two years, we would do all the work for free.

> ### How to Support Local Farmers
> Maybe we've convinced you about the importance of local, sustainable farming but you can't farm yourself (or not yet). Here are some ideas on how you can support your local farmers:
>
> Shop at your farmers' market and/or subscribe to a CSA.
>
> Volunteer at a farm.
>
> Sit on the board of your local market.
>
> Ask the produce manager in your grocery store to stock local food, and to label the origin of all produce.
>
> If you subscribe to a CSA, get more involved. If you have skills like website design or organizing, offer them.
>
> Grow a little food garden in a visible place.
>
> Dig up your lawn and plant food, then tell people why it's important.
>
> Invite a friend over when you're cooking rutabaga, celeriac, squash, and parsnips in the winter. Start a conversation about seasonality.
>
> Choose local apples instead of papaya and mango.
>
> Thank your farmer.
>
> Get involved with urban agriculture or food security organizations.
>
> Teach your sons and daughters to cook.

Melanie, our L.J. crop manager, poses during strawberry harvest. Our favourite all-around variety is Seascape but Tri-star is unsurpassed for flavour.

This was a natural inclination for us because on our own farms, we had never tracked our hours. In retrospect, this may have been the only strategy we could have used to build the company—the attitude of faith supported us in what we were doing and gave us the time we needed to learn. Now, when we're hiring, we have the experience of having done the jobs ourselves, and are able to direct other people.

In the context of large conventional farms that are propped up by annual subsidies, and small farms that rely on free labour from family and travelling volunteers, it was revolutionary to us that farming could be done in a way that paid everyone involved a decent wage. In order to warrant a full-time greenhouse manager's time, the project we still call L.J. spilled out of the greenhouse to encompass field crops. We hoped that along with their regular salad orders, chefs might be tempted to buy our other offerings.

We now have a quarter-acre in strawberries and a half-acre in annual vegetables, as well as a rotation of salad greens and nightshades, such as peppers and eggplant, in the greenhouse. In essence, L.J. started with a ten-year jump on each of our own farms because we were bringing our knowledge and passion to the project. More than just a section of the farm, Long John is an experiment in collective growing, and a culmination of our years of experience.

Hiring Employees

As Saanich Organics grows, we've found ourselves in the unexpected position of being employers. It's strange being on the other side of the fence. Part of our motivation in running our own businesses was to avoid working for "the man." So now, we're learning how to run a successful business without becoming "the man." We

Caleb brings the harvest into the washing area. Bathtubs are used for washing produce, a cement pad keeps our feet dry, and the Rubbermaid bins drain in our homemade rack.

in checking her references and interviewing her, were all about interpersonal skills and work ethic. Instead of "Is she fun to work with?" we should have asked, "Is she detail-oriented, and how are her organizational skills?" We wasted a month and caused her a lot of grief before we realized we had the perfect person in the absolutely wrong job. Once we sorted this out, we had a passionate, self-directed, efficient fieldhand who worked with us for two years. We'd jump at the chance to work with her again.

Sometimes we become too close with employees. The season begins, and some workers are either slow or unreliable, or both. They are almost always urban young people who have little experience with hard, manual labour where speed is of the essence. Of these people, some catch on to the work quickly, but others don't, and the problem is magnified by both practicalities and emotions. Mid-season, it is difficult to find the time to rehire, so we hold out hope that the employees will improve. We are also empathetic and emotive. We care about our employees, and when we see them in challenging personal and/ or work situations, it becomes more and more difficult to fire them. Conversely, we have also had the deeply satisfying experience of seeing people completely fulfilled and even transformed by their experience on the farm. It is hard, but we know we need to fire those who aren't working out sooner, so we can give the opportunity to the right person. As Heather said to Robin, "You've got to channel the 'man brain'—is she doing the job or not?" Perhaps this problem goes all the way back to the hiring process, and we continue to tweak our method to try to filter out slow, flaky, or overly needy people.

have an advantage in that not so long ago, each of us worked at low-wage jobs, and we are mindful of the cost of living and of the need to create a meaningful and respectful work atmosphere. We strive to give our employees responsibility, to share our passion, and to allow them to participate as fully as possible in the joy of farming. One challenge has been to realize that we can't expect the same level of emotional investment from an employee that we ourselves have in our farms and in Saanich Organics.

Another challenge is placing the right person in the right job. We hired a farmer acquaintance and fell in love with her passion for the work; her job was to do both field work and administrative work. We assumed that because she was smart, hard-working, and eager, she could do anything. We needed a computer-savvy, organized, personable administrator, but the questions we asked,

We wish we could say that after several years of hiring employees and apprentices, we know what we're doing. However, the truth is that we're still surprised sometimes. Occasionally we'll hire someone who appears solid: confident, tough, hard-working, committed to ecological agriculture, and passionate, only to have them quit a few weeks in. More often, we're surprised the other way. One spring, a box customer, Karlyn, called to inquire about apprenticing opportunities. She came out to meet us, and we decided on a rotating schedule where she'd volunteer on all our farms. After each of us met her, we compared notes, and all said things like, "Well, she *said* all the right things" and "She's not looking for a paid job, so we don't have anything to lose." None of the three of us thought she'd stick around long. For one thing, there was her appearance: she was tall, slender, and urban-chic. Not only that, but she had makeup on! And as Lamont said, "She smells too good to be a farmer." We also thought her plans were unrealistic. She had rearranged her life to require less money so that she could volunteer nearly full-time while maintaining a couple of shifts a week at her lucrative pub job. We expected that she'd burn out, realize she needed more money, and/or get bored when the work got monotonous and difficult.

Well, from the first day she walked onto the farm until the bitter end of the season, Karlyn was a lifesaver. Not only was she competent and hard-working, she was also one of those people who made everything she touched easier. Whenever a situation seemed difficult, she was the one to say, "Why don't I just . . ." and the problem was solved. Karlyn definitely helped us see, and question, our first impressions.

Dumpster Diving

We tend to hire a very eclectic crew of farmhands. The folks who are drawn to working hard for modest wages are with us because organic farming resonates with their values. It is inspiring to meet our new farmhands each year and to learn about their path to organic agriculture. Some folks are cerebral and they come to us from the academic stream. Others are from activist circles and they bring with them some extreme views that challenge us and expand the way we see things.

Over the years there have been several dumpster divers. They are anarchists who believe in alternative food systems as a means to bring justice to the world. Some of them are heavily involved with food-reclamation projects all over the city (visiting dumpsters and redistributing food that would otherwise go to waste). Our organic and local distribution resonated with them and they were loyal, hard-working employees. We respected their mission and were happy to have them use our excess produce. One of these individuals, however, presented an interesting problem: he stank! This is not an understatement.

We at Saanich Organics have never been accused of excessive hygiene. Much to the contrary, we're usually "*au naturel*." But this guy could make us back away from ten feet or more. He rarely did laundry, so the smell was literally garbage juice soaked into his clothes that was getting warmed by his sweating body under the sun. Our poor manager, who volunteered to give folks a ride to the farm, had to grin and bear it with the windows down all the way to town and back. It's hilarious in retrospect, but at the time it was a hardship.

The farm crew gathers for lunch and camaraderie. Left to right: Melanie, Ali, Dennis, Jeremy, and Andrea.

More than once, we have made the mistake of assuming that the same things that motivate us motivate our employees. We are gradually learning that some people want more direction rather than more responsibility, and that asking people what they want is easier than trying to guess. We are learning that we need to give all our employees a vision of the big picture, which involves the sometimes-monumental task of freeing up time for meetings.

At a workshop on employee relations, the presenter suggested that employers look around their staff rooms to make sure they are warm and inviting. Robin and Heather looked at each other in shame as they realized that our employees didn't even have chairs to sit on to eat their lunch, unless someone pulled up a wheelbarrow. Now they have a covered area with table and chairs, a hand-washing sink, and a message board. We've also improved our system for distributing excess produce among the staff. We make a more conscious effort to express our appreciation for their important role, with occasional surprise gifts and a year-end feast.

Our employees bring more than their labour to the farm: they bring their diverse personalities, their enthusiastic energy, their new perspectives, and their refreshing sense of fun. We have momentum in the community that helps bring in new employees. Our current employees are invaluable for sending good people our way, and returning volunteers and apprentices have also become valued employees.

The Finances

Saanich Organics makes money in three ways: from the box program, commercial sales, and Long John. Although we also sell at the farmers' markets together, those sales go straight to our own farms, not through the business accounts. We want to make it clear that our farms are our primary sources of income. The efficient marketing provided by the Saanich Organics structure makes this possible. Restaurants form the bulk of Saanich Organics' business, followed by our box program, the farmers' markets, and a few stores. We like this diversity of outlets because it gives us some security. As consumer trends change, some markets will become more or less important, but we won't lose all our markets at once.

Our sales through the box program and the commercial division are marked up to cover our costs and make a modest profit. We went into this business because we wanted to be farmers. We want the farmers who sell to Saanich

Organics to get a fair return on their produce, so we offer them a good price and keep our markup as low as we can. That being said, we also value our time. We want to make sure that the business does not lose money, that it makes a modest return to compensate us for the many hours we spend packing orders, planning, hiring, and taking care of a million details. Our contentment with modest profits and slow growth is contrary to conventional business wisdom. Most distributors aim to buy low and sell high. We aim to buy high, and sell slightly higher.

Indeed, more business-oriented friends and acquaintances have often expressed amazement when they realize we have a waiting list for our box program and that our chefs can't all get all the produce they want. "Why don't you expand by buying from more farmers or importing produce?" they ask. Above all, we are farmers first. We want to spend our time growing food, so profiting from the marketing end of the business has not been a priority. The distribution is necessary, and we certainly want it to be as efficient as possible, but our goal is to make our primary income from growing produce, rather than from distributing it.

Saanich Organics' biggest expense is the cost of the produce that we buy from our own farms and the other farms that sell with us. We mark up the produce that we sell through the box program by 30 per cent. Produce that goes to commercial sales is marked up by 15 per cent. The difference reflects the additional work involved in organizing and packing the residential boxes. These markups pay for administration, bookkeeping, delivery, help in the packing room, rental of the garage, office supplies, and hydro.

Activism in Action

We feel that the main reason we are so lucky in finding great employees who are interesting, driven, inspired, and committed is that we provide an opportunity for them to live their values and participate in something they believe in. They often have marketable skills, like a university education, and other opportunities are available to them, but they choose to work with us, and we are very grateful. We pay as well as we can, but it is still a subsistence wage. It is an important indicator of the value of local, organic agriculture, that some of our best and brightest are rolling up their sleeves and diving in.

Melanie can't contain her excitement with her harvest of arugula seed. We've been thrilled with the extent of our success at seed-saving in our first year, despite being a wet and challenging year.

Seed Saving

Our latest endeavour within Saanich Organics is saving our own seed. Historically, seed-saving was a necessity for a farmer but we learned to farm without knowing the basics of harvesting seed for our own use. Our complete dependence on seed companies has begun to feel uncomfortable. The more we learn about food politics, the more we crave a connection to all parts of our food production. Rachel was shocked to read that her favourite hybrid tomato, Big Beef, is owned by Monsanto (she doesn't grow that one anymore!). Robin is thoroughly inspired by her work with USC Canada, a seed sovereignty NGO, from which she learned about small farmers building resilience to climate change by breeding their own bio-regionally adapted seed. Heather has been saving seed from her cucumbers and discovered that she created a different and more productive variety that works better than others on her farm. For all these reasons (and perhaps because seed is getting really expensive), we decided to start Seeds of the Revolution, our own seed company. We've started small, with a half-dozen varieties, but we intend to expand. This new direction will diversify our farm products, as well as give us the freedom to opt out of the industrial system bit by bit.

Long John makes money by selling produce to Saanich Organics as if it were a farm on its own. Long John's biggest expense is labour, followed by soil amendments and equipment. Our starting wage for farmhands in 2010 was $10.50 per hour (the minimum wage in BC is $8.75 per hour). Returning employees get raises. Each year, Long John gives us grey hairs by going deep into debt, sometimes by as much as $8,000, before sales finally surpass expenses around August.

Another major expense is delivery. Tim earns $3.25 per residential delivery, which is paid by the customers. For commercial sales, we pay him 4.25 per cent of the gross sales and we recover about half of this cost by charging the clients $5.00 per delivery. We used to have a complicated formula for paying Tim, which included gas surcharge, hourly wage, and a per-stop fee. The percentage system is a vast improvement because his earnings are directly proportional to his work. (See Appendix C for Saanich Organics' Income Statement from 2008.)

Personal Compatibility

Our ability to work together and to share a vision for the future is the main reason that Saanich Organics works. We share the dream of making a living from farming. At a workshop on business management, we learned a classification tool that divides people into four personality types: analytics, drivers, expressives, and amiables. According to this scheme, a functional business requires all four types. We identified ourselves as three of the four personality profiles, while each of us has elements of the fourth. This may help to explain why we work so well together.

Sitting around the fire pit at Three Oaks Farm. We enjoy our impromptu gatherings to share a meal, or just to take a break from the busy-ness of the business.

Rachel is an analytic. She is logical and she takes the time to understand all the elements that contribute to a problem. She believes it is important to do things right, and is emotionally reserved until decisions are made. Heather is a driver. She is practical, efficient, and task-oriented. When she has a vision, she can see the steps that need to be taken and can delegate so that everyone feels involved and inspired. Robin is an expressive. She is an idea-generator and has energy for expansion and change. She enjoys setting the course, and then looking to the others for guidance on how to make it happen. All three of us have components of the amiable character type. We are relationship-oriented, empathetic, and able to see other viewpoints.

Another factor in our compatibility is our stage of life. We met when we were struggling to make our new businesses work. We were all working hard, but felt overwhelmed by the many different tasks involved in both growing and marketing, and we hoped that working together would make our businesses more efficient. Heather and Rachel dreamed of having children, and they were looking for ways to run a farm with a family. We were all looking for the quality of life we wanted out of farming, and were open to change. We all had a similar desire for financial stability but were content to follow a cautious growth strategy. Nine years later, we have matured together and still share a similar outlook.

Our co-operation has evolved over a decade. When we started working together, none of us had any idea how Saanich Organics would evolve. We have good communication, we love to work hard, and we trust each other's intentions. We have also been incredibly lucky to be surrounded by a nurturing and talented community. We inspire and challenge one another; we support each other and celebrate our successes together.

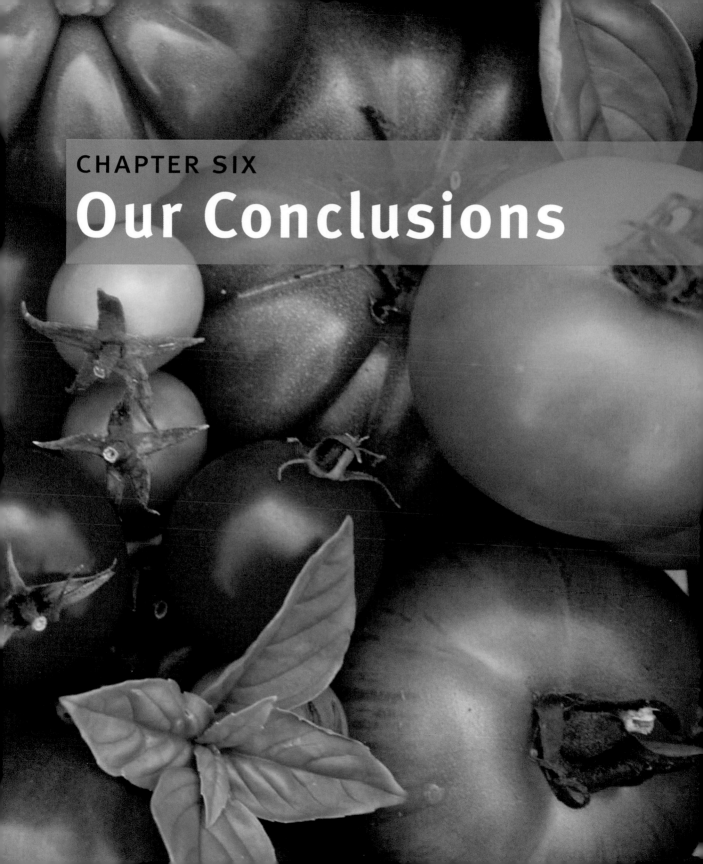

CHAPTER SIX

Our Conclusions

Through our years of farming, much has changed. Our farms now reliably provide us with an income. Each of our farms has seasonal staff and/or WWOOFers and apprentices, so our farming community continues to grow. Several times each season we are reminded of the many friendships and allegiances we've built over the years. Surprise visits at the market and e-mails from across the miles trigger fond memories and make us realize how much impact our little business has had. Not only are we better at what we do, but just as importantly, our soil improves year after year, so our farms are significantly more productive. Our markets expand and we improve the efficiency of Saanich Organics.

However, there are some things that don't change. Each year, during the busiest season, we still feel overwhelmed and even occasionally question our decision to farm. In spite of our best efforts, we still have crop failures, which remind us of the importance of diversity! Greens ravaged by slugs or flea beetles, or an early frost on our butternut squash is heartbreaking, but because we have twenty-five other crops, it doesn't take long before the disaster becomes yet another funny story.

When we talked about why we are still so deeply committed to farming, each of us kept returning to this question: what we would do if we weren't farming?

Rachel: What else would I do? What other work could I find that could be so fulfilling, that would both satisfy and reward me?

Robin: My work is tangible. Other people think farmers are at the mercy of forces they can't control, but I feel otherwise. I can see my crops, I can decide what has to be done, and I'm my own boss. Doing physical work is therapeutic. Farming is sometimes physically exhausting, but I find it mentally invigorating. I don't play music, I don't paint, I don't scrapbook. My creativity is my farm.

Heather: What am I going to do, put on nylons and go to an office? That just doesn't feel like an option anymore.

For Rachel and Heather, farming is closely tied to how they want to raise their children. Both of them often wish that they had more time with their kids, that they weren't always so rushed, and had more time for music, crafts, or other activities. However, they both love that their kids can engage with them in meaningful, practical work and that they don't have to choose between being a stay-at-home mom or a working mom with children in daycare.

Our farms are financially successful by our own estimation, and certainly compared with national statistics of farming incomes. However, making a full-time living from farming continues

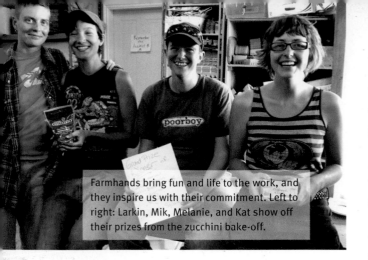
Farmhands bring fun and life to the work, and they inspire us with their commitment. Left to right: Larkin, Mik, Melanie, and Kat show off their prizes from the zucchini bake-off.

to be a challenge. In writing this book, we kept returning to questions of finances. All three of us have been somewhat surprised to discover in ourselves an entrepreneurial spirit. We each have a drive to make money, but continue to find that we don't make as much as we feel we deserve for the amount of work we put in.

Acknowledging our entrepreneurial drive has been complicated for each of us. We all see huge problems with our capitalist, consumerist culture and each of us values many things (environment, social justice, family, friends) above money. We see that in many industries, money is made at the *expense* of the environment. Compensation for labour is not the norm in our culture. Professional athletes make millions for entertaining us, while those who grow the food we all need for our survival are paid less than minimum wage. Even within agriculture, there is a paradox: large farms are considered valuable and small farms insignificant, or "hobbies," but small farms actually produce more food and make more money per acre. Small farms create more jobs and are generally more ecologically sound. It's time that their value was recognized.

Robin: I bought into the fallacy that I wasn't deserving of money. I used to believe that if I loved

what I was doing, I shouldn't get paid for it. Work is work, and fun is fun. If you're farming because it's your love, you shouldn't be making a living.

Rachel: When I started farming, I didn't need to make much money, and I didn't. Later, when I started setting financial goals, I started reaching them.

Heather: I get angry when people talk about the difficulty of making money farming and then say, "But it's a lifestyle decision," as if that explains away or excuses our culture's neglect of the needs of those who feed us. Working for an oil company or becoming a doctor are also "lifestyle decisions" and no one expects those people to work for free.

Whatever our thoughts about our money-obsessed, consumer culture, the fact remains that we both need and deserve to be paid for our work. There is no earthly reason why farmers should have to hold off-farm jobs to "support the farming habit," as a neighbour says.

Work, play, and home are all melded together in the life of a farmer. Our lives are not compartmentalized; we don't leave our work at the office. It's not unusual to walk in on Heather in the kitchen, feeding her children while freezing berries for the winter boxes and taking a call from a chef. This is both a blessing and a curse; sometimes the inescapable nature of our work is overwhelming. However, most days the work is so much of who we are that we can't distance ourselves from it. This holistic approach to life is something that we've all brought to the business of Saanich Organics. Our work brings us fun, companionship, support, exercise, renewal, and inspiration, so we don't have to seek those elements elsewhere; the fact that we don't often have

Karen shows off her harvest of leeks. Karen has since moved away to a new farm but remains a dear friend, and a trusted consultant for our seed business.

time for recreation or entertainment isn't such a hardship. All our parties may include some business talk, but often our work together feels a bit like a party.

Each November we have a Saanich Organics year-end party. We prepare a feast with food we have grown, and share it with everyone who has been involved with the business that year and their families (employees, contractors, farmers who have sold with us). The last several years, chefs we work with have generously donated their time to prepare this meal. The menu usually looks something like this:

    Green salad
    Celeriac salad
    Roasted beets
    Carrot/parsnip soup
    Cabbage, black bean, tomato, and corn salad
    Chicken we raised

    Wild, locally caught salmon
    Roasted butternut squash
    Local bread with basil/sundried tomato pâté
    Butternut squash/brandy pie
    Apple pie

We have a very deep feeling of satisfaction and pride, standing in a house full of good friends and looking at all the food that we produced—from seed to plate. Late at night, after one such gathering we received this e-mail from our dear friend and farmhand, Karen:

"hey women i was thinkin of you all as i drove home along that long and windy road and here is what i was thinkin it was amazing tonight to see all the people that you have brought together through your business very cool and i just wanted to take a moment to make sure you take a moment to feel good about such a crowd i was talking to rachel later in the evening talking how amazing it is to have so many young farmers in this community . . . so many more than when rachel first started here and this is no coincidence you are part of this . . . if you look around the room so many of these people are here and farming becuz of your support and openness and encouragement not to mention land and tools so i just wanted to pass along appreciations for you all i'm lucky to have hooked up with you and hope to continue in the next season to work together and if you ever felt like you weren't an activist . . . inspiring young farmers and creating opportunities for fair wage, support and land is as cuttin edge as it gets yeah to on-farm activism and thanks for another season of timez with you and your dogs and your kids and your soil."

Writing *All the Dirt* has been satisfying for all of us. We've enjoyed remembering our first years, and each winter, our work on this book has sparked our enthusiasm for the next spring planting season. We've come to know each other better and have had a chance to reflect on how, and more importantly why, we farm. We hope you've enjoyed our stories and learned a bit from our experiences. If you do decide to find a piece of land and plant some seeds, we wish you success, however you define it. We hope that your crops will grow, that you will nurture your soil and yourself, and that you will be able to build a strong, supportive community around you.

Resources

Books

Ableman, Michael. *On Good Land: The Autobiography of an Urban Farm*. San Francisco: Chronicle Books, 1998.

———. *Fields of Plenty: A Farmer's Journey in Search of Real Food and the People Who Grow It*. San Francisco: Chronicle Books, 2005.

Berry, Wendell. *The Unsettling of America: Culture and Agriculture*. San Francisco: Sierra Club Books, 1977.

Coleman, Eliot. *The New Organic Grower: A Master's Manual of Tools and Techniques for the Home and Market Gardener*. Halifax: Nimbus Publishing, 1995.

———. *Four Season Harvest*. White River Junction, VT: Chelsea Green Publishing, 1999.

———. *The Winter Harvest Handbook: Year-Round Vegetable Production Using Deep Organic Techniques and Unheated Greenhouses*. White River Junction, VT: Chelsea Green Publishing, 2009.

Desmarais, Annette A. *La Via Campesina: Globalization and the Power of Peasants*. Halifax: Fernwood Publishing, 2007.

Diamond, Jared M. *Collapse: How Societies Choose to Fail or Succeed*. Los Angeles: Viking Press, 2005.

Fern, Marshall Bradley. *Rodale's All New Encyclopedia of Organic Gardening: The Indispensable Resource for Every Gardener*. Vermont: Rodale Press, 1993.

Gershuny, Grace, and Joe Smillie. *The Soul of Soil: A Soil Building Guide for Master Gardeners and Farmers*. White River Junction, VT: Chelsea Green Publishing, 1999.

Gilkeson, Linda. *Natural Insect, Weed and Disease Control*. Indiana: Trafford Publishing, 2006.

Herriot, Carolyn. *The Zero Mile Diet: A year round guide to growing organic food*. Madeira Park, BC: Harbour Publishing, 2010.

Heyasson, D.G. *The Vegetable Expert*. London: Expert, 1997.

Howard, Ronald J., J. Allen Garland, and W. Lloyd Seaman. *Diseases and Pests of Vegetable Crops in Canada*. The Canadian Phytopathological Society and The Entomological Society of Canada, 1994.

Jason, Dan. *The Whole Organic Foods Cookbook: Safe, Healthy Harvest from your Garden to your Plate*. Vancouver, BC: Raincoast Books, 2001.

Jeavons, John. *How to Grow More Vegetables Than you ever Thought Possible on Less Land Than you can Imagine*. San Francisco: Ten Speed Press, 1979.

Kingsolver, Barbara. *Animal, Vegetable, Miracle: A Year of Food Life*. Toronto: HarperCollins Publishers, 2007.

Lappe, Frances Moore. *Diet for a Small Planet*. New York: Ballantine Books, 1971.

Matthias, Laura. *ExtraVeganZa: Original Recipes from Phoenix Organic Farm*. Gabriola Island, BC: New Society Publishers, 2006.

Mollison, Bill. *Introduction to Permaculture*. Tasmania: Tagari Press, 1991 (reprinted 2002).

———. *Permaculture: A Designers' Manual*. Tasmania: Tagari Press, 1997.

Nestle, Marion. *Food Politics: How the Food Industry Influences Nutrition and Health*. Berkeley: University of California Press, 2002.

Patel, Raj. *Stuffed and Starved: Markets, Power, and the Hidden Battle for the World's Food System*. Toronto: HarperCollins Publishers, 2007.

Pollan, Michael. *The Omnivore's Dilemma: A Natural History of Four Meals*. New York: Penguin, 2006.

———. *In Defense of Food: An Eater's Manifesto*. New York: Penguin, 2008.

Salatin, Joel. *You Can Farm: The Entrepreneur's Guide to Start and Succeed in a Farming Enterprise*. White River Junction, VT: Chelsea Green Publishing, 1998.

———. *Pastured Poultry Profit$*. Swoope, VA: Polyface Press, 1996.

———. *Everything I Want To Do Is Illegal: War Stories From the Local Food Front*. Swoope, VA: Polyface Press, 2007.

Schlosser, Eric. *Fast Food Nation: The Dark Side of the All-American Meal*. Toronto: Houghton Mifflin, 2001.

Shiva, Vandana. *Stolen Harvest: The Hijacking of the Global Food Supply*. Cambridge, MA: South End Press, 2000.

Smith, Alisa, and J.B. MacKinnon. *The 100 Mile Diet: A Year of Local Eating*. Toronto: Random House, 2007.

Solomon, Steve. *Growing Vegetables West of the Cascades*. Seattle: Sasquatch Books, 1989.

Theriault, Frederic, and Daniel Brisebois. Cog Practical Skills Handbooks, *Crop Planning For Organic Vegetable Growers*. Ottawa, ON: Canadian Organic Growers Inc., 2010.

Films

El Contrato

Food, Inc.

Fresh: New Ideas on what we Eat

From Seed to Seed

The Future of Food (futureoffood.com)

Island on the Edge (dvcuisine.com/dvds/island-on-the-edge)

King Corn (kingcorn.net)

Living Lightly

Outstanding in her Field, 1994. (youtube.com/watch?v=ibO
    BBSO9M5k&feature=related)

The Real Dirt on Farmer John (angelicorganics.com/therealdirt/)

Seeds of Change (seedsofchangefilm.org)

SuperSize Me (rottentomatoes.com/m/super_size_me/)

Magazines

Acres USA

BC Organic Grower

The Canadian Organic Grower

Small Farm Canada

Small Farmers Journal

Websites We Like

acornorganic.org

almfarms.org

attra.ncat.org

bcseeds.org

certifiedorganic.bc.ca

cog.ca

craftfarmapprentice.com

fairviewgardens.org

farmbase.ca

farmerswithoutborders.org

fermetournesol.qc.ca

ffcf.bc.ca

fooddemocracy.org

foodsecurecanada.org

ggfarm.com

globallifestyles.ca

haliburtonfarm.org

helpx.net

saanichorganics.com

sare.org

saskorganic.com

scytheworks.ca

sfp.ucdavis.edu

soilapprenticeships.org

usc-canada.org

wwoof.org

Appendix A: Farming Set-Ups Heather's Spreadsheets: Planting Data

CODE	CROP	VARIETY	SOURCE	DATE PURCHASED	DATE PLANTED	BLOCK/DIRECT	QUANTITY	BED LENGTH	GERM COMMENTS	TRANS PLANT DATE	SPACING	DESCRIPTION	COMMENTS	AMMENDMENTS	More comments
Merg-501	Loganberries		Deacon Vale Farm	26-Feb-05	1-Mar-05		25	100'							
CF3-501	Carrots	Bolero	West Coast	2003	6-Mar-05	Direct	1/3 CF				hand seed		dug morning glory first	Generous Dolopril	
CF3-502	Radishes	Altaglobe	West Coast	2004	6-Mar-05	Direct	1/8 CF				hand seed		dug morning glory first	Generous Dolopril	
Moo1-501	Snap peas	Super Sugar Snap	William Dam	2005	8-Mar-05	Direct	2 rows	50'			seeder plate	eaten, slow, fill-planted Apr 19	Tilled, raked 1-2 weeks ago	2 okara, 2 Dolopril	
Moo2-501	Snap peas	Super Sugar Snap	William Dam	2005	8-Mar-05	Direct	2 rows	50'			seeder plate		Tilled, raked 1-2 weeks ago	2 okara, 2 Dolopril	
Jup3-501	Lettuce	Paris Island Cos	Full Circle	2000	9-Mar-05	blocks	96		barely any germ	29-Apr-05	12", 4 rows		seeded thickly	approx 6 okara, 1/3 bag lime, 1 scoop compost	
Jup3-501	Lettuce	Sierra MI	William Dam	2005	9-Mar-05	blocks	96			29-Apr-05	12", 4 rows	good!!	disc, cult.lots, tilled in ammendments	approx 6 okara, 1/3 bag lime, 1 scoop compost	
Jup3-501	Lettuce	Envy	Johnny's	2003	9-Mar-05	blocks	96			29-Apr-05	12", 4 rows	good!!	disc, cult.lots, tilled in ammendments	approx 6 okara, 1/3 bag lime, 1 scoop compost	
Jup3-501	Lettuce	Rave	Johnny's	2003	9-Mar-05	blocks	96			29-Apr-05	12", 4 rows	good!!	disc, cult.lots, tilled in ammendments	approx 6 okara, 1/3 bag lime, 1 scoop compost	
Jup3-501	Lettuce	Esmeralda	West Coast	2001	9-Mar-05	blocks	48			29-Apr-05	12", 4 rows	good!!	disc, cult.lots, tilled in ammendments	approx 6 okara, 1/3 bag lime, 1 scoop compost	
Jup3-501	Lettuce	gourmet lettuce mix	West Coast	2004	29-Apr-05	Direct	1/3 bed, 4 rows		good		seeder plate		disc, cult.lots, tilled in ammendments	approx 6 okara, 1/3 bag lime, 1 scoop compost	filled in end of bed
	Celery	Ventura	West Coast	2005	9-Mar-05	blocks	432								
Ear8-501	Radishes	Altaglobe	West Coast	2004	13-Mar-05	Direct	4 rows	60'			seeder plate		tilled, disc, cultivate, rake	1/3 bag lime, 4wb compost	
Ear7-501	Spinach	Olympia/Spargo	WC/WD	03 and 05	13-Mar-05	Direct	cross-rows	50'			seeder plate		tilled, disc, cultivate, till	1/3 bag lime, 4wb compost, 5 okara	
Ven1-501	Beets	Red Ace	West Coast	2005	14-Mar-05	Direct	4 rows	75'			seeder plate		tilled 3 times over 2 weeks	4 okara, 1/5 bag lime	
Moo9-501	Radishes	French Breakfast	West Coast	2005	14-Mar-05	Direct	4 rows	50'			seeder plate		cultivate, disc, tilled	2/5 bag lime	
Moo10-501	Radishes	French Breakfast	West Coast	2005	14-Mar-05	Direct	4 rows	50'			seeder plate		cultivate, disc, tilled	2/5 bag lime	
	Tomatoes	Black Prince	Farmgirl	2005	16-Mar-05	blocks	17								
	Tomatoes	Green Zebra	Farmgirl	2005	16-Mar-05	blocks	17								
Jup10-501	Tomatoes	Brandywine	Johnny's	2003	16-Mar-05	blocks	24	1/3 of 100'		26-May-05	2', 2 rows		disc, cult, till, rotovate, amend each hole	shovel compost, tiny sulpomag, light dolopril, sprinkle kelp flour each hole	lots left over
CF3-504	Tomatoes	Brandywine	Johnny's	2003	16-Mar-05	blocks	24	middle of CF		27-May-05	2', 3 rows		weeded, amended each hole	shovel compost, tiny sulpomag, light dolopril, sprinkle kelp flour each hole	lots left over
Jup10-501	Tomatoes	Moskvich	Johnny's	2003	16-Mar-05	blocks	24	1/3 of 100'		26-May-05	2', 2 rows		disc, cult, till, rotovate, amend each hole	shovel compost, tiny sulpomag, light dolopril, sprinkle kelp flour each hole	lots left over
CF3-504	Tomatoes	Moskvich	Johnny's	2003	16-Mar-05	blocks	24	edge of CF		27-May-05	2', 1 row		weeded, amended each hole	shovel compost, tiny sulpomag, light dolopril, sprinkle kelp flour each hole	lots left over
Jup10-501	Tomatoes	Viva Italia	William Dam	2005	16-Mar-05	blocks	24	1/3 of 100'		26-May-05	2', 2 rows		disc, cult, till, rotovate, amend each hole	shovel compost, tiny sulpomag, light dolopril, sprinkle kelp flour each hole	lots left over

Code	Crop	Variety	Source	Year	Seed Date	Method	Number	Length		Transplant	Spacing		Bed Prep	Amendments	Notes
CF3-504	Tomatoes	Viva Italia	William Dam	2005	16-Mar-05	blocks	24	edge of CF		27-May-05	2', 1 row		weeded, amended each hole	shovel compost, tiny sulpomag, light dolopril, sprinkle kelp flour each hole	lots left over
Jup6-501	Parsley	Dark Green Italian	West Coast	2003	16-Mar-05	blocks	432	100'	TERRIBLE	13-May-05	12", 4 rows		disc, cult lots, rotovate, tilled in amendments	1 scoop compost, dolopril	All the seedlings only filled 3/4 of this bed (bad germ),
Jup6-501	Parsley	Forest Green	West Coast	2003	16-Mar-05	blocks	432	100'		13-May-05	12", 4 rows		disc, cult lots, rotovate, tilled in amendments	1 scoop compost, dolopril	All the seedlings only filled 3/4 of this bed (bad germ),
Jup6-501	Parsley	Dark Green Italian	West Coast	2003	13-May-05	Direct	1/4 bed, 4 rows	100'	TERRIBLE		hand seed		disc, cult lots, rotovate, tilled in amendments	1 scoop compost, dolopril	
Jup6-501	Pepper	Early Jalapeno	West Coast	2002	16-Mar-05	blocks	14						disc, cult lots, rotovate, tilled in amendments	1 scoop compost, dolopril	filled in end of bed
Jup14/5-501	Chard	Rhubarb	West Coast	2005	20-Mar-05	blocks	240			29-Apr-05	18", 3 rows	sm, but good!	disc, cult.lots, tilled in ammendments	approx 6 okara, 1/3 bag lime, 1 scoop compost	
Jup16-501	Chard	Rhubarb	West Coast	2005	20-Mar-05	blocks	240			29-Apr-05	18", 3 rows	sm, but good!	disc, cult.lots, tilled in ammendments	approx 6 okara, 1/3 bag lime, 1 scoop compost	
Jup12-501	Chard	Rhubarb	West Coast	2005	20-Mar-05	blocks	240			3-May-05	1/2 bed 18", 3 rows	sm, but good!	disc, cult.lots, tilled in ammendments	4 okara, 1/3 bag lime, 1 scoop compost	
Ura3-501	Chard/Kale	Assorted	Assorted		20/26 Mar	blocks				12-May-05	18", 3 rows		disc, cult bit, rotovate, tilled in amendments	4 okara, compost, dolopril	Used up leftover seedlings, mostly Fordhook Chard and Kale
Jup12-501	Chard	Perpetual	William Dam	2005	20-Mar-05	blocks	120			3-May-05	1/2 bed 18", 3 rows	sm, but good	disc, cult.lots, tilled in ammendments	4 okara, 1/3 bag lime, 1 scoop compost	
Ear6-501	Salad	Assorted	WC/WD	2004/2005	21-Mar-05	Direct	cross-rows	50'			hand seed		brassicas and beets only	4 okara, 3wb compost, 1/4 bag lime	
Moo8-501	Spinach	Mazurka	West Coast	2005	21-Mar-05	Direct	cross-rows	50'			hand seed			3 okara, 2wb compost, 1/5 bag lime	
CF3-503	Salad	gourmet lettuce mix	West Coast	2004	23-Mar-05	Direct	1/5 CF				hand seed		pulled old salad, forked.	okara, dolopril	
Jup15-501	Kale	Red Russian	William Dam	2005	26-Mar-05	blocks	240	100'		29-Apr-05	18", 3 rows	sm, but good	disc, cult.lots, tilled in ammendments	approx 6 okara, 1/3 bag lime, 1 scoop compost	
Jup13-501	Kale	Red Russian	William Dam	2005	26-Mar-05	blocks	240	100'		3-May-05	1/2 bed 18", 3 rows		disc, cult.lots, tilled in ammendments	4 okara, 1/3 bag lime, 1 scoop compost	
Jup11-501	Kale	Nero Di Toscana	William Dam	2005	26-Mar-05	blocks	240	100'		3-May-05	1/2 bed 18", 3 rows		disc, cult.lots, tilled in ammendments	4 okara, 1/3 bag lime, 1 scoop compost	
Jup11-501	Kale	Improved Siberian	West Coast	2005	26-Mar-05	blocks	240	100'		3-May-05	1/2 bed 18", 3 rows		disc, cult.lots, tilled in ammendments	4 okara, 1/3 bag lime, 1 scoop compost	
Jup13-501	Kale	Improved Siberian	West Coast	2005	26-Mar-05	blocks	240	100'		3-May-05	1/2 bed 18", 3 rows		disc, cult.lots, tilled in ammendments	4 okara, 1/3 bag lime, 1 scoop compost	
	Eggplant	Hybrid	Johnny's	2005	30-Mar-05	blocks	96								
	Eggplant	Dusky	West Coast	2005	30-Mar-05	blocks	48								
	Eggplant	Fairy Tale	West Coast	2005	30-Mar-05	blocks	48								
	Pepper	Ancho	West Coast	2003	30-Mar-05	blocks	48								
CF4-501	Salad	gourmet lettuce mix	West Coast	2004/2005	31-Mar-05	Direct	1/3 CF				hand seed				
	New Zealand Spinach		West Coast	2002	1-Apr-05	blocks	240						soaked seeds		
Ura4-501	Lettuce	Envy	Johnny's	2003	1-Apr-05	blocks	144			12-May-05	12", 4 rows		cult, disc, rotovate, tilled in amendments	4 okara, 1yd sea soil, bit dolopril	
Ura4-501	Lettuce	Rave	Johnny's	2003	1-Apr-05	blocks	96			12-May-05	12", 4 rows		cult, disc, rotovate, tilled in amendments	4 okara, 1yd sea soil, bit dolopril	

DATE	CODE	CROP	VARIETY	SOLD LB	SOLD UNITS	MARKET	PRICE PER	TOTAL	RECEIPTS	COMMENTS
2-May-05	Moo-9-501	radishes	French Breakfast	5			2.50		209552	
9-May-05	Moo-10-501	radishes	French Breakfast	14		C.D.	2.50		209552	pulled up bed, last harvest
9-May-05	Moo-8-501	spinach	French Breakfast	4		C.D.	8.50		209552	
9-May-05	Ven-1-501	beet thinnings	Red Ace	4		C.D.	2.80		209552	
16-May-05	Moo8-501	spinach	Mazurka	5		C.D.	8.50		209552	
2-May-05	Ear8-501	radishes	Altaglobe	13.5		C.D.	2.50	33.75	489502	LOTS more
2-May-05	Ear6-501	Salad		5		C.D.	10.00	50.00	489502	
2-May-05	CF4-501	Salad		2		C.D.	10.00	20.00	489502	
2-May-05	Mer2-401	Rhubarb		18		C.D.	1.50	27.00	489502	
10-May-05	Ear8-501	radishes	Altaglobe	2		C.D.	2.50	5.00	489502	
10-May-05	CF3-503	Salad		3		C.D.	10.00	30.03	489502	
10-May-05	CF4-501	Salad		5		C.D.	10.00	50.00	489502	
10-May-05	Mer3-401	Rhubarb		10.5		C.D.	1.50	15.75	489502	
10-May-05		eggs			10	C.D.	4.00	40.00	489502	
10-May-05		Chard		3		C.D.	2.80	8.40	489502	overwinter
17-May-05	Ear6-501	Salad		8		C.D.	10.00	80.00	489502	
17-May-05	CF4-501	Salad		6		C.D.	10.00	60.00	489502	
17-May-05	CF3-503	Salad		1		C.D.	10.00	10.00	489502	
17-May-05	CF3-503	Braising Mix		2		C.D.	4.00	8.00	489502	
17-May-05		eggs			8	C.D.	4.00	32.00	489502	
7-May-05	Ear8-501	radishes	Altaglobe		7	Moss St	1.75	12.25	489503	
7-May-05	Mer5-401	Artichokes			5	Moss St	1.00	5.00	489503	
7-May-05	Ear6-501	Salad			6	Moss St	3.00	18.00	489503	
30-May-05	Moo8-501	spinach	Mazurka	2		C.D.	6.00		209553	
30-May-05	Ven-1-501	baby beets	Red Ace			C.D.	2.50		209553	
20-May-05	Ear6-501	Salad		2		C.D.	10.00	20.00	489502	
20-May-05	CF4-501	Salad		2		C.D.	10.00	20.00	489502	
20-May-05	Ear6-501	Braising Mix		3		C.D.	4.00	12.00	489502	
24-May-05	Jup3-501	Salad		2		C.D.	10.00	20.00	489502	lettuce
24-May-05		d.beans		9.5		Box	4.90	46.55	489505	last year's

Date	Code	Product	Variety		Qty	Location	Price	Amount	Ref	Notes
24-May-05		eggs			12	Box	4.00	48.00	489505	
31-May-05	Jup15-501	Kale	red russian	9		Box	2.30	20.70	489505	
31-May-05	Jup3-501	Salad		0.5		Box	10.00	5.00	489505	
31-May-05		eggs			3	Box	4.00	12.00	489505	
21-May-05	Ear7-501	Spinach	Spargo		6	Moss St	5.00	30.00	489507	1/2 lb bags
21-May-05	Mer4-401	Artichokes			21	Moss St	1.00	21.00	489507	
21-May-05		S. Squash	Zucchini		1	Moss St	2.00	2.00	489507	
28-May-05	Cf4-502	Arugula			16	Moss St	3.00	48.00	489508	100g bags
28-May-05	Jup3-501	Lettuce	assorted		16	Moss St	2.50	40.00	489508	
28-May-05	Cf4-501	Lettuce			17	Moss St	3.00	51.00	489508	150g bags, cut
31-May-05	Jup3-501	Salad		3.5		C.D.	10.00	35.00	489510	
31-May-05	Jup3-501	Lettuce			107	C.D.	1.40	149.80	489510	
31-May-05	Mer4-401	Artichokes			33	C.D.	1.00	33.00	489510	
31-May-05		eggs			8	C.D.	4.00	32.00	489510	
4-Jun-05	Mer5-401	Artichokes			7	Moss St	1.00	7.00	489513	
4-Jun-05	Jup12-501	Chard	Rhubarb		6	Moss St	2.00	12.00	489513	
4-Jun-05	Jup14-501	Chard	Rhubarb		6	Moss St	2.00	12.00	489513	
4-Jun-05	Jup13-501	Kale	Improved Siberian		3	Moss St	2.00	6.00	489513	
4-Jun-05	Jup15-501	Kale	red russian		6	Moss St	2.00	12.00	489513	
4-Jun-05	Jup3-501	Lettuce	assorted		16	Moss St	2.50	40.00	489513	
4-Jun-05	Jup5-501	spinach	Spargo		11	Moss St	3.00	33.00	489513	
4-Jun-05	Jup4-501	radishes	Easter Egg		15	Moss St	1.75	26.25	489513	
4-Jun-05	Jup3-501	Salad	gourmet lettuce mix		4	Moss St	3.00	12.00	489513	
4-Jun-05	Ura5-501	Salad	assorted		5	Moss St	3.00	15.00	489513	

The first page of the spreadsheet includes all the planting data. These are the column headings with brief explanations:

Code: This tells me what bed and what year the crop came from.

Crop: What kind of vegetable.

Variety: I have to keep track of what varieties I'm growing to learn which ones work for me.

Source: Where I got the seed (or transplants).

Date Purchased: What year the seed was purchased (for audit trail, and also because this can teach me why a crop didn't germinate, and how long seed lasts).

Date Planted: This is essential for learning when to plant different crops in your microclimate.

Block/Direct: This tells me whether the crop was started in blocks, pots, or direct-seeded.

Quantity: If they are started in blocks, I record the number of starts; if direct-seeded, then the number of rows.

Bed Length: Together with spacing info, and quantity info, this tells me how much of the crop went into the ground. This forms part of the audit trail for certification, and also lets me calculate income per square foot for each crop.

Germination Comments: General observations on germination. Together with planting date, seed source, and age of seeds, this gradually teaches me how and when to plant different crops.

Transplant: This is the date of transplanting (left blank for direct-seeded crops).

Spacing: This forms part of the audit trail, and also is very handy to look at each spring when I'm asking myself, "Oh gosh, now how far apart do I usually plant these?"

Description: A general observation at the time of transplanting.

Comments: This is where I record the soil preparation.

Amendments: Here I record what I add to the bed. This goes along with the "inputs sheet" as part of the records required for certification.

More comments: This is where I explain anything that is not clear from all the other columns.

The second spreadsheet contains harvesting data. This includes both amounts harvested (necessary for audit control summary in certification) and prices (to make income calculations simple). These are the column headings, with explanations:

Date: Date of harvest.

Code: This ties the information together with the planting data page. The code tells me where in the field the crop was planted.

Crop: Occasionally, this will say something different from the corresponding code on the planting data sheet. For example, I may record that I planted kale on the first sheet, but if I sell part of that crop as "sprouting brassicas" in the spring, that is what I would enter in the crop column.

Variety: Self-explanatory.

Sold lb, or Sold Units: For each entry, I fill in only one of these columns. For example, if I sell bulk chard to a restaurant, it is entered under "Sold lb," but if I bunch it for the market, it is entered under "Sold units."

Market: This is where I sold the produce. Over the years, my markets have simplified as I sell everything either at the Moss Street

Community Market, to the Saanich Organics box program, or to Saanich Organics commercial division.

Price Per: This price can be the price per pound, or per bunch or head, whatever the case may be.

Total: In this column I enter a formula so Excel automatically multiplies the amount sold by the price.

Receipts: This is the invoice number, so the information is tied to the hard copy that the customer received.

Comments: This can be anything from the size of the bunch to the appearance of the crop, or anything else.

In the winter I look in detail at all this data, and make a summary sheet of yields of each crop. It is also informative to note first and last harvest dates of each crop.

I also keep a spreadsheet for farm expenses. After all the receipts are entered, I can sort them by category and see where I'm spending money. I can also separate out the capital expenses that are depreciated over several years from the expenses that are written off on that year's taxes. All vehicle expenses are grouped together, and I write off a portion of those costs (depending on the percentage of total annual vehicle use that was farm-related).

These are the column headings for my expense spreadsheet:

Date: Self-explanatory.

Receipt number: This helps me find the original receipt if I need to. (I write this number on the original receipt.)

Company: i.e., Integrity, West Coast, et cetera.

Code: Our tax person wants this (she gave us the codes she wants).

Description: i.e., feed, seeds, Sea Soil, et cetera.

GST: Keeping this amount separate makes it easy to claim back.

Total Cost: Amount paid including tax.

Cost: This is the amount without GST, the amount we deduct at tax time.

Division: For my own interest, i.e., "chickens," "veggies," "fencing," et cetera.

Robin's Lease Agreement

Here is a guide for discussion between potential landlords and tenant farmers:

1. Length of Lease: It can take five to ten years to get land productive enough to be profitable, so negotiate a lease for as long as possible. We settled on five years, with a year-to-year renewal after that. We negotiated two hundred days' notice if either my landlords or I had to break the lease. This would ensure that I had time to harvest crops and the landlords to recruit another farmer.

2. Capital Expenditures: The land I leased was a fenced horse pasture with no buildings. Since the landlords already had a glass greenhouse that they wanted to put up, and a garden shed that was on another property, we negotiated that they would supply the materials and I would supply the labour. I worked with a carpenter to put up both of these structures. It made sense to us that the landlords paid for permanent infrastructure improvements that made their land more appealing to prospective tenants.

I had to extend the fence upward to make it deer-proof, and again they supplied the netting and two-by-fours so I could secure my one-acre plot. In the ensuing years, I have installed drainage. I paid for the backhoe work and most of the materials, but they gave me some pipes and parts that they had from other projects. I have erected other greenhouses on the property and paid for them myself, but I intend to take them with me if I leave. I have built some temporary structures out of pallet wood that are functional but not attractive. The landlords are fine with this, but the understanding is that I would disassemble these structures if I leave.

3. Rent and Utilities: I did a small survey on the price of leased land in our area, and the average was forty dollars per acre per month, so that's the rate we decided on. I have been fortunate that my landlords haven't raised the price of my lease during the twelve years of my occupancy. There is a separate meter for my utilities, to operate the pump and the power for my potting shed (lights, freezer, plug in for power tools, et cetera). There is a spring on the property, so I use that water for irrigation. I have city water for washing the veggies, and this has a separate meter where it branches from the main source of water for their house. I pay for all the utilities and water that I use.

4. Water: I can't stress this one enough. The farm needs to be guaranteed access to water at all times. If there's any question of not having enough during certain times of the day or year, you really need to think it through. I feel grateful to have two sources of water on the property—spring water and city water—so I have unlimited pressure and volume. Lack of water causes so much stress that it may not be worth entering an agreement if water is limited.

5. Dispute Resolution: We wrote into the lease that if we couldn't come to an agreement, we would seek counselling from a dispute-resolution centre. We found one in the phone book and decided that it would be suitable should we need

some outside help. I think it is a good idea to have a strategy in writing for a worst-case scenario.

6. Other Considerations

a. Traffic on the Land: My landlords have been really accommodating about extra traffic on the property. Our lease has a clause about permission for operating a farm stand and holding some educational events on the land. I have two or three staff parking places at the top of the driveway, and when I have farm tours or a plant sale, sometimes the parking is chaotic. It is good to think through the boundaries and carrying capacity of the land, and what type of events may cause friction. It is nice to have the option of being creative with direct marketing.

b. Chemical Drift: I don't have to worry about chemical drift from my landlords but this can be a consideration if you are trying to maintain organic status and there is conventional gardening or production happening on the property. Maintaining buffer zones is crucial, so you may need to ask for a guarantee that chemicals won't be used within a prescribed distance of your production area.

c. Farm Animals: I found out that although my landlords were tolerant of a rooster, the neighbours were not, and for the sake of neighbourhood harmony, my rooster had to go. I had been keen to breed my own meat birds but I just had to abandon that plan. It is good to talk in detail about what is and isn't acceptable in terms of livestock so you can think through your options on the land.

d. Aesthetics: It is important to talk about what a farm may look like: e.g., Reemay cloth, pallet-wood construction, compost piles, and harvesting bins. If the landlords have a low tolerance for shabby infrastructure, they may be willing to pay for better materials. It may be worthwhile taking your prospective landlords on a tour of a friend's farm, just to give them a sense of what it might look like.

e. Selling your Farming Business: I may be in the situation in a few years of feeling ready for a bigger parcel of land. The easiest scenario for me would be to sell my farm business and start afresh on a new parcel. Since I have so many apprentices who are starting out, it is very likely that one of them could be a good fit on the farm and would appreciate all the improvements I've made. I didn't negotiate anything like this with my landlords but I will probably be doing this in future. I think it's good to have an exit clause whereby you could have the potential to recoup some of the expenses, and the landlords can screen the potential new candidates.

Map of Feisty Field Spring/Early Summer 2006
Arugula in ground for 4 months 750 sq. ft. @ $4.00 per sq. ft.
Beets, Red Ace, successions in ground for 6 months 600 sq. ft. @ $4.25 per sq. ft.
Carrots in ground for 3 months 900 sq. ft. @ $5.40 per sq. ft.
Celeriac in ground for 5–8 months, not harvested until fall
Cucumbers in small cold frame for 3 months 300 sq. ft. @ $2.00 per sq. ft.
Failed Cutting Lettuce 600 sq. ft. @ $0.40 per sq. ft.
Garlic in ground for 9 months 900 sq. ft. @ $2.30 per sq. ft.
Greenhouse Salad Greens in production for 10 months 1500 sq. ft. @ $6 per sq. ft.
Japanese Turnips, successions in ground for 6 months 300 sq. ft. @ $2.50 per sq. ft.
Leeks, in ground for 5–9 months, not harvested until fall
Lettuce in ground for 4 months 150 sq. ft. @ $4.00 per sq. ft.
Lettuce, successions in ground cycling for 5 months 900 sq. ft. @ $4.00 per sq. ft.
Pac Choi in ground for 2 months 3000 sq. ft. @ $2.80 per sq. ft.
Peas in ground for 4 months 300 sq. ft. @ $1.20 per sq. ft.
Potatoes in ground for 4 months 600 sq. ft. @ $1.16 per sq. ft.
Red Turnips in ground for 2 months 300 sq. ft. @ $2.50 per sq. ft.
Salad Greens, successions in ground for 5 months 1200 sq. ft. @ $4.00 per sq. ft.
Salsify in ground for 6 months, not harvested until frost
Strawberries in ground for 2 years 900 sq. ft. @ $3.10 per sq. ft.
Swiss Chard in ground for 11 months 600 sq. ft. @ $1.20 per sq. ft.
Winter Squash in ground for 7 months 900 sq. ft., not harvested until fall

These field maps are a bed-by-bed representation of Robin's crops for each growing season. The crop value is included in the season when it is harvested even though some crops are taking space in the field for a longer period. There is not enough information in these maps to truly help with crop planning because so many factors determine profitability (i.e. insects, fertility problems, marketing failure, adverse weather), but they do provide a snapshot of crop returns for a given season.

Map of Feisty Field Summer/Late Fall 2006
Arugula in ground for 3 months 600 sq. ft. @ $4.00 per sq. ft.
Basil in ground for 3 months 300 sq. ft. @ $3.30 per sq. ft.
Beets, Red Ace, successions in ground for 6 months 300 sq. ft.
Carrots in ground for 5 months 900 sq. ft. @ $5.40 per sq. ft.
Celeriac in ground for 5–8 months 900 sq. ft. @ $5.20 per sq. ft.
Greenhouse Salad Greens in production for 10 months 1500 sq. ft. @ $6.00 per sq. ft.
Japanese Turnips, successions in ground for 6 months 300 sq. ft.
Kale in ground for 12 months 600 sq. ft.
Leeks, in ground for 5–9 months 600 sq. ft. @ $4.50 per sq. ft.
Lettuce in ground for 4 months 150 sq. ft.
Lettuce, successions in ground cycling for 5 months 900 sq. ft. @ $4.00 per sq. ft.
Overwintering onions in ground for 9 months 900 sq. ft. @ $2.30 per sq. ft.
Purple Top Turnips in ground for 3 months 900 sq. ft. @ $2.80 per sq. ft.
Salad Greens in small cold frame for 3 months 300 sq. ft. @ $4.00 per sq. ft.
Salad Greens, successions in ground for 5 months 1200 sq. ft. @ $4.00 per sq. ft.
Salsify in ground for 6 months 200 sq. ft. @ $4.20 per sq. ft.
Strawberries in ground for 2 years 900 sq. ft.
Swiss Chard in ground for 11 months 600 sq. ft.
Winter Squash in ground for 7 months 900 sq. ft. @ $1.20 per sq. ft.

Appendix B: Saanich Organics' Organization

Saanich Organics: Suppliers and Outlets

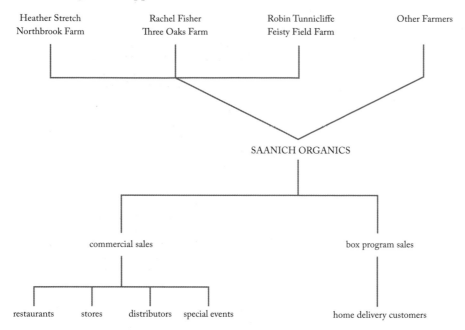

Heather Stretch
Northbrook Farm

Rachel Fisher
Three Oaks Farm

Robin Tunnicliffe
Feisty Field Farm

Other Farmers

SAANICH ORGANICS

commercial sales

box program sales

restaurants   stores   distributors   special events

home delivery customers

Saanich Organics: Internal Structure

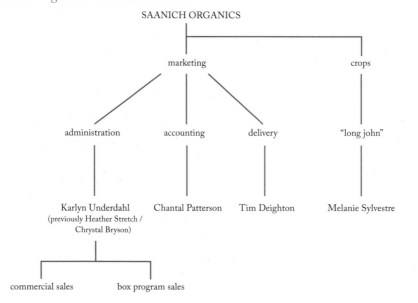

SAANICH ORGANICS

marketing

crops

administration   accounting   delivery

"long john"

Karlyn Underdahl
(previously Heather Stretch /
Chrystal Bryson)

Chantal Patterson   Tim Deighton

Melanie Sylvestre

commercial sales   box program sales

Big Taste from Small Farms

For three evenings in a row, I've walked past Heather's apprentices Jill and John, with lawn chairs pulled up beside the chicken pen. They are like new parents fawning over this first batch of wee chicks. One of the greatest parts of having apprentices is that they give you fresh eyes for all the wonderful things on the farm.

Honey Peppered Turnips

1 tablespoon butter
2 tablespoon(s) honey
1 pound turnips, peeled, 1/4 inch dice
1/2 teaspoon black pepper
kosher salt
parsley, chopped

Melt the butter with the honey in a medium saucepan over medium-low heat. Stir in the turnips and pepper. Cook, covered, until tender, about 12 minutes. Add salt to taste and sprinkle with parsley.

Beet and Parsley Salad

1 bunch baby beets without greens
1 cup packed fresh flat-leaf parsley leaves
1/4 teaspoon salt, or to taste
1/4 teaspoon sugar, or to taste
1/8 teaspoon black pepper
2 teaspoons extra-virgin olive oil
2 teaspoons balsamic vinegar

Trim beets, and then cut into quarters. Steam until tender. Toss beets with parsley, salt, sugar, and pepper in a serving bowl until sugar is dissolved. Add oil and toss to coat. Sprinkle vinegar on salad and toss again. Serve immediately. Sprouts are also good in this salad.

**1438 Mt Newton Cross Road
Saanichton, BC V8M 1S2
250-818-5807**

For the week of June 29, 2010

Baby Beets	Lettuce
Beet Tops	Parsley
Garlic Scapes	Spinach and Turnips
Kale	Strawberries

Spinach Strawberry Salad

1 bag spinach, rinsed and torn into bite-size pieces
4 cups sliced strawberries
1/2 cup vegetable oil
1/4 cup white wine vinegar
1/2 cup white sugar
1/4 teaspoon paprika
2 tablespoons sesame seeds
1 tablespoon poppy seeds
In a large bowl, toss together the spinach and strawberries. In a medium bowl, whisk together the oil, vinegar, sugar, paprika, sesame seeds, and poppy seeds. Pour over the spinach and strawberries, and toss to coat.

Sprout Humus

1 cup sprouts
1 Tbs. tahini
1 Tbs. lemon juice
1 Tsp. olive oil
1 clove garlic (more or less - to taste)
1 tsp. ground cumin
1/2 tsp. salt
1/2 tsp. ground white pepper

Blend all ingredients in a food processor and serve.

News From the Farm

We're sad to report that the melon plants are not looking good, nor are the cucumbers, that are in the same family (cucurbit). Cucurbits don't like wet feet, and the soggy soil from the cold spring has made their lives miserable, unless they were taken first by slugs. Heather has just re-seeded cukes for the third time and is quite worried because she has been saving seed for her little white cukes for many years now and may lose her stock if this last batch doesn't take. Our fingers are crossed for a long sunny autumn season.

info@saanichorganics.com

Saanich Organics: Sample Sales Spreadsheet

Offered — 13-Oct-09

Robin	Long John	Rachel	Heather	Chrystal & Illya	Melanie	F1	Offered	Ordered	Shortage/Surplus	Veggies	Variety	Comments
U	U	U	U	U	U	U						
-	-	-	50	-	-		50	51	1	eggplant	white/purple mix	
-	-	-	5	-	-		5	5	-	cabbage	danish ball head	
-	-	3	5	-	-		8	3	(5)	arugula	larger	
7	5	-	3	-	-		15	8	(7)	arugula	salad size	salad alm
30	-	70	-	20	-		120	20	(100)	beets	red	red ace
10	-	5	-	-	-		15	13	(2)	beets	baby	
-	-	20	20	-	-		40	2	(38)	beets	winterkeeper	
-	-	1	-	-	-		1	5		parsnips		
-	-	40	-	-	-		40	32	(8)	rutabaga		
-	-	-	40	-	-		40	8	(33)	dandelion		
5	20	20	30	-	-		75	57	(18)	Braising Mix		
50	-	-	-	-	-		50	52		celeriac		
-	-	-	-	-	100		100	70	(30)	sunflower sprou	(bagged) transit mix	
50	30	-	50	50	-		180	73	(107)	carrots	nantes, d	
-	-	10	-	-	-		10	20		carrots	baby rainbow	
-	-	30	-	-	-		30	22	(8)	carrots	rainbow	
-	-	-	5	-	-		5	9		chard	rainbow	
-	-	5	-	-	-		5	9		chard	red	red
-	-	5	-	-	-		5	9		chard	green	green
-	-	-	-	100	-		100	13	(87)	pea shoots	(bagged) transitional	
-	30	-	-	-	-		30	20	(10)	turnips	purple top	
-	20	-	10	-	-		30	22	(8)	turnips	baby japanese	
-	-	10	-	-	-		10	15		radish	watermelon	
-	-	-	20	-	-		20	24		radish	purple winter	
-	-	32	-	-	-		32	32	-	radish	black spanish	
-	6	-	-	-	-		6	6		kale	redbor	
-	30	-	-	-	-		30	4	(26)	daikon	small	
10	-	-	-	-	-		10	13		ching chiang	choi	
20	-	-	-	-	-		20	10	(10)	Cilantro		
-	5	20	100	-	-		125	127		parsley	italian	italian, by
-	60	-	-	-	-		60	76		parsley	curly	
65	-	-	-	-	-		65	60	(5)	pac choi		
-	5	5	-	-	-		10	10		broccoli		
-	150	5	70	-	-		225	91	(134)	sweet pepper	mix	
-	-	2	10	-	-		12	7	(5)	hot pepper	mix	
-	-	-	-	5	-		5	6		tomatillos		
30	5	10	10	8	2		65	72		salad		
3	-	2	4	-	-		9	13		spinach		
-	-	-	20	-	-		20	20	-	tomatoes	heirloom	sugar, sw
-	-	20	-	-	-		20	-	(20)	tomatoes	green	
12	-	-	-	-	-		12	12		lettuce	red leaf	buttercup
-	10	-	-	-	-		10	13		mei quing choi		spagetti 4
30	-	20	-	-	-		50	42	(8)	leeks		red kuri h
40	40	20	5	-	-		105	40	(65)	collards	green	sweet du
-	-	-	30	-	-		30	-	(30)	collards	purple	sugar loa
-	-	-	-	50	50		50	66		cucumber	long english	blue hubb
-	-	-	-	50	50		50	2	(48)	cucumber	american slicing	
-	15	-	-	-			15	4	(11)	fennel		
4	-	-	-	-	-		4	4	-	green onions	green	n
30	-	20	50	30	-		130	2	(128)	w. squash	acorn (asst)	overripe
-	-	200	-	-	-		200	10	(190)	w. squash	heirloom varietie	p
-	200	100	20	-	30		350	31	(319)	w. squash	delicata/sweet d	nutty red
60	200	-	20	-	200		480	150	(330)	w. squash	red kuri/b.nut/sp	sm, dry b
-	-	-	12	-	-		12	18		strawberries	pints	
-	-	-	-	-	-		20	81		pears	8 oz boxes	u
272	**150**	**398**	**347**	**458**	**82**	**250**						

Ordered — 13-Oct-09

Veggies	Box	Share	Planet Organic	Stage	La Piola	Cridge Centre	Olympic View	Lucy's	Oak Bay Marina	Café Brio	Haro's	Paprika	Sips	Devour	Heron Rock	Spinnakers	Spud	Rebar	Lifestyle Market	TOTAL
(call/tue)	call	call	fri	d	fri	d	call	tue	call	tue	fri		tues	call	call	tue	call	tue	fri	d	tues	call
eggplant	30	9				3					5		2					4		51
cabbage							2	1					2							5
arugula										3										3
arugula							1	2							5					8
beets						10					10									20
beets								3								10				13
beets														2						2
parsnips								5												5
rutabaga									7	10						15				32
dandelion		3															2	3		8
Braising Mix		37			2		2			2	10						4			57
celeriac						10	3	2	5				2		10	20				52
sunflower sprou	60						2	4					2		2					70
carrots	60							7										6		73
carrots															20					20
carrots						5		10			5		2							22
chard		6					3													9
chard		6									3									9
chard		6									3									9
pea shoots					3					4				2				4		13
turnips									7	10						3				20
turnips							2									20				22
radish							1	1			2							1		15
radish	20						1	2										1		24
radish	30						1											1		32
kale		6																		6
daikon							2							2						4
ching chiang																10		3		13
Cilantro						8				2										10
parsley		119																8		127
parsley		58																	18	76
pac choi	60																			60
broccoli											5	5								10
sweet pepper		90												1						91
hot pepper	4							2										1		7
tomatillos															6					6
salad					5	4		2	3	2		15	6	2	8	10	15			72
spinach							2			10								1		13
tomatoes								10							10					20
tomatoes																				-
lettuce						12														12
mei quing choi					4		2			5				2						13
leeks							5						15	2			20			42
collards		6														20	8		6	40
collards																				0
cucumber	60					6														66
cucumber							2													2
fennel														4						4
green onions					4															4
w. squash							2													2
w. squash															10					10
w. squash					15	2									10	4				31
w. squash	150																			150
strawberries						6					6	6								18
pears	60										1							20		81
TOTAL	334	313	33		30	36	31	46	69	7	57	17	15	8	41	161	22	16	33	

Saanich Organics: Sample Weekly Invoice

Saanich Organics Purchase Order — Robin — 13-Oct-09

Saanich Organics
1438 Mt.Newton X Rd.
Saanichton, BC V8M 1S1
250-818-5807

info@saanichorganics.com

What you are billing for this week

No.	What	Specifically	Qty Sold To Restaurants	Price	Sold To Restaurants	Qty Sold To BOX	Price	Sold To BOX	Total
1	salad size	arugula	7	$10.00	$70.00	-	$10.00	$ -	70.00
2	red	beets	10	$2.25	$22.50	-	$2.25	$ -	22.50
3	baby	beets	10	$3.00	$30.00	-	$3.00	$ -	30.00
4	0	Braising Mix	4	$4.75	$19.00	-	$4.75	$ -	19.00
5	0	celeriac	52	$2.50	$130.00	-	$2.50	$ -	130.00
6	choi	ching chiang	13	$3.05	$39.65	-	$3.05	$ -	39.65
7	0	Cilantro	10	$1.00	$10.00	-	$1.00	$ -	10.00
8	0	pac choi	0	$2.10	-	60	$2.10	$126.00	126.00
9	0	salad	30	$10.00	$300.00	-	$10.00	$ -	300.00
10	0	spinach	4	$8.75	$35.00	-	$8.75	$ -	35.00
11	red leaf	lettuce	12	$2.00	$24.00	-	$2.00	$ -	24.00
12	0	leeks	22	$2.50	$55.00	-	$2.50	$ -	55.00
13	green	green onions	4	$3.20	$12.80	-	$3.20	$ -	12.80
14	acorn (asst)	w. squash	0	$1.55	$ -	-	$1.55	$ -	-
15	red kuri/b.nut/spagh	w. squash	0	$1.55	$ -	-	$1.55	$ -	-

Total sold to RESTAURANTS — $747.95
Total sold to BOX — $126.00
Grand Total — $873.95

What This Farm Should Pack Up To Sell

TOTAL AVAILABLE FROM THIS FARM	TOTAL BOUGHT FROM THIS FARM	TOTAL UNSOLD FROM THIS FARM	Box	Share	Planet Organic	Stage	La Piola	Cridge Centre	Olympic View	Lucy's	Oak Bay Marina	Café Brio	Haro's	Paprika	Sips	Devour	Heron Rock	Spinnakers	Spud	Rebar	Lifestyle Market	TOTAL RESIDENTIAL
7	7	-														2				5		0
30	10	20							10													0
10	10	-							10													0
4	4	1				4						0		0	0		0					0
50	52	(2)							10		3	2	5			2		10	20			0
10	13	(3)						8										10		3		0
20	10	10							8			2										0
65	60	5	60																			60
30	30	-					5	4		2	3	2			14		0	0	0	0		0
3	4	(1)						2							2						0	0
12	12	-						12														0
30	22	8					5								15			2				0
4	4	-					4															0
30	0	30																				0
60	0	60																				0
	238																					

Oly and Lucy's celeriac count

Saanich Organics			
Income Statement			
January 1, 2008 to December 31, 2008		2008	
		Total	
Box - Revenue		53,276.50	
- Special Orders		72.50	
- Eggs		341.25	
- Annual Box Charge		510.00	
- Delivery Charges		6,581.00	
		60,781.25	
Box- Cost of Goods Sold			
- Produce		40,743.73	
- Equip Rental		675.00	
- Bad Debts		143.75	
- Accounting		2,463.75	
- Supplies		1,541.79	
- Admin Charges		3,061.69	
- Donations		391.50	
- Delivery Charges		6,613.85	
		55,635.06	
Box - Net Income		**5,146.19**	
Commercial - Revenue		185,148.15	
- Delivery		3,920.00	
Commercial - Cost of Goods Sold			
- Produce		160,331.12	
- Bad Debts		153.15	
- Equip Rental		675.00	
- Supplies		686.43	
- Accounting		2,899.00	
- Admin Time		8,061.60	
- Delivery Charges		8,079.05	
		180,885.35	
Comm - Net Income		**8,182.80**	
Greenhouse (L J) - Revenue		33,285.76	
Greenhouse - Expenses			
Greenhouse Supplies		7,026.51	
Greenhouse Equip Rental		261.00	
Depreciation		1,377.48	
Manager Hours		8,365.32	
Labour		13,929.00	
		30,959.31	
Greenhouse (L J)- Net Income		**2,326.45**	

Saanich Organics			
Income Statement			
January 1, 2008 to December 31, 2008		2008	
		Total	
Other Income		818.14	
Other Expenses			
Advertising		1,221.66	
Bank Charges		370.02	
Repairs - Computer		37.50	
Miscellaneous		532.22	
Telephone		902.28	
Spoiled Produce		270.63	
Office Supplies		242.62	
Water & Hydro		567.00	
		4,143.93	
Net Income (Loss)		**9,304.35**	

This annual income statement represents an average year. In our best year, the business netted $18,000. In our worst, and the only year when we ran a loss, we lost $4,000. Take note that this income is just that of our marketing business and we consider it a bonus. Our main income is made from our farms. The statement is organized into: Box program revenue and expenses, followed by net income; Commercial sales revenue and expenses, followed by net income; Long John (greenhouse) revenue and expenses, followed by net income; other income and expenses; and finally, Total net income for the year. (Some less important lines have been omitted to keep the statement uncluttered.)

Index

Acknowledgments

We have a whole new understanding of the term "labour of love" after writing this book. As if being farmers wasn't hard enough, we had to go and write about it too! It could not have happened without the loving support of Heather and Rachel's partners, who minded our farms while we were away on writing retreats, who put the kids to bed while we spent long hours at the computer, and who brought us coffee and chocolate to keep us going. Robin has been the glue that kept this project going by keeping the end vision in sight, planning island getaways for writing, and being ever-eager to have this book ready to present at the next conference.

The seed of the idea to write a book was first planted by Lara Fisher during a meeting with Robin and Rachel. Heather was away in North Carolina at the time visiting Lamont's family; she confided later that when she and Lamont heard the news that we were writing a book, they thought Robin must have overdosed on her thyroid medication! Our thanks to Lara for getting the ball rolling.

We are indebted to Catherine Etmanski for her gentle but thorough editing of our rough draft, which improved it immensely; to Lucy Mei Lee for the title of the book; and to Ireta Fisher for the use of her Mayne Island cottage for a writing retreat. We also want to thank the photographers among us who contributed many of the beautiful photos, most notably Jill Banting, Jesse Cottingham, and Madeleine Gauthier.

We would also like to acknowledge the people who make our farms viable: our box program subscribers, our market customers, and the chefs who order produce week in and week out. Special gratitude, of course, to those lovely chefs who have created giant feasts from our produce for our year-end parties: Brock, Louis and Genevieve, Mohammed, and Sheena.

Thank you to all those in our organic farming community, especially the IOPA and its members. Your spirit of co-operation rather than competition has made this community thrive and made us all better people, as well as better farmers.

We'd especially like to extend our thanks to Ruth Linka and the TouchWood publishing team. Your recognition and transformation of our diamond in the rough, and commitment to manifesting our vision in book form has made the process so meaningful and satisfying.

Last, but not least, we'd like to thank the people who have gotten their hands dirty making our farms something worth writing about. We hesitate to try to name everyone, because we'll inevitably miss someone important, but here's our best shot (with apologies in advance for omissions): Cyra, Wendy, Chrystal, Ilya, Larkin, Jeremy, Ali, Aida, Ian, Hannah, Dara, Kate, Alex S., Melanie, Karen, Julie, Scott, Alex M., Jessie J., Jesse H., Sasha, Lindsay, Mik, Dennis, Pat (Mom), Gord (Dad), Matthew, Jenny, Scott L., Devin, Aaron, Jennie, Karlyn, Tim, Cat, Brenda, Gracie, Barb, Chantal, Jen, Joshua, John, Jill, Carly, Jesse C., Vanessa, Kat, Kate, Briar, Hannah, Steve, Kira, Liz, Carla, Justin, Peter, Randi, Catherine E., Andrew, Caleb, Nadia, Sachiko, Jesse, Catherine L., Paula, Rachelle, Paul, Chris, Ken, and dear Angie, who has appeared at the Moss Street Community Market every week to help us set up our booth for so many years that we have lost track.